Seeing the Light

Seeing the Light

Regaining Control of Our Electricity System

David Morris

Institute for Local Self-Reliance
Minneapolis, MN   Washington, DC

The New Rules Project identifies and promotes policies that nurture healthy communities and strong local economies. The project is a comprehensive effort to articulate a new vision of politics and economics for the twenty-first century. The New Rules Project is a program of the Institute for Local Self-Reliance, a 25-year-old nonprofit research and educational organization that provides techical assistance and information on environmentally sound economic development strategies.

For more information about the New Rules Project, including a list of other titles, please visit our website at www.newrules.org or contact us at:

Institute for Local Self-Reliance

1313 5th Street S.E.
Minneapolis, MN 55414
Phone: 612-379-3815
Fax: 612-379-3920

2425 18th Street NW
Washington, DC 20009
Phone: 202-232-4108
Fax: 202-332-0463

www.ilsr.org

Printed in the United States of America

First Edition

Illustrations by Ken Avidor
Book design by Holle Brian

ISBN 0-917582-88-8

05 04 03 02   5 4 3 2 1

Oh subtle fire, soul of the world
beneficent electricity
You fill the air, the earth, the sea
The sky and its immensity.

—*C.J.M. Barbaroux* (1784)

Scarcely anything of the world before electrification has remained untouched: how things work, how and where work gets done, how people are transported, how food is cooked and served, how people keep in touch…what they see. The very smell of cities has been altered.

—Electric Power Research Institute,

EPRI Journal (1979)

Contents

Illustrations

Foreword

America, with about 4.5 percent of the world's population, uses about 30 percent of the world's electricity. Electric utilities account for more of the nation's total capital investment than any other industry.

Since the New Deal dismantled the early, huge, controversial electric utility holding companies, the generation of electric power has been considered a sleepy sector. The growth of the industry and its rate of return were subjected to broad state and federal regulation. Consequently, it became a classic "widows and orphans" investment—throwing off regular dividends without much risk or controversy.

That is not to imply that utility regulation was without major problems. Many utility commissions were captured by the companies they were supposed to regulate. Monumentally stupid decisions were sometimes made—including various nuclear turkeys that nearly bankrupted their owners. But most consumers received dependable service at reasonable rates, and most shareholders received reasonable dividends on secure investments.

All that is now changing. The wave of deregulation that transmogrified the savings and loan industry, the telecommunications industry, the airline industry, etc., is now sweeping through the electric utility sector. Twenty-four states have introduced retail competition to date, and many others have it on the agenda.

The repercussions will be far-reaching. The deregulation of electricity will affect the distribution of wealth, the quality and reliability of service, the speed of the transition to renewable power sources, the drive for increased efficiency, etc. Whether produced by coal, oil, nuclear, or large hydro power plants, electricity lies at the root of many of the nation's most serious environmental problems. Deregulation can either promote or retard their solution.

It will also either promote or constrain the use of a new generation of small, decentralized power generation technologies. These are poised to replace traditional economies of scale with new economies of mass production—much as inexpensive PCs are replacing powerful mainframe computers. These technological choices could have important implications for community control.

The public discussion accompanying these changes has been relatively mute. It is another example of what I call barbell policy debates. At one end is a glut of arcane articles in specialized journals written by policy wonks for their peers. At the other extreme is a raft of simplistic, focus-group-tested slogans that reduce complex choices to bumper strips. The ground in between—thoughtful, accessible essays designed to provoke serious thinking by intelligent readers—has been surprisingly thin, considering the money and power involved in the decisions.

With *Seeing the Light*, David Morris moves into this void. David digests the complex history of electricity in American to provide a concise context for our current choices. Then he proposes a set of understandable "new rules" to guide society through the monumental chess game of deregulation.

In any social experiment of this size and complexity, the only sure bet is that every policy will produce unanticipated consequences. *Seeing the Light* does a splendid job of anticipating some likely consequences that other have missed.

Seeing the Light is a lucid introduction to the things we should look out for. An hour or two spent with this book will provide any reader with the background to participate knowledgeably in one of the most far-reaching policy debates of our time.

Denis Hayes
President of the Bullitt Foundation
and former director of the federal
Solar Energy Research Institute

Preface

The prospect of being hanged in the morning, English writer Samuel Johnson counselled two hundred years ago, "concentrates the mind wonderfully." The rolling blackouts and spiraling electricity prices in California in early 2001 have concentrated the nation's mind on our electricity system and, more specifically, on the deregulation feeding frenzy that swept through the country between 1996 and 2000.

Among the 24 states (plus the District of Columbia) that had approved deregulation plans, at least seven announced by mid February 2001 that they are delaying implementation: Nevada, Arkansas, New Mexico, Montana, West Virginia, North Carolina and Oklahoma.

For the nation as a whole, this is good news. The passion for electricity deregulation was never embraced by the average citizen. A choice of electricity supplier was far, far down on their wish-list for an electricity system. A handful of large industrial customers and a handful of independent power suppliers defined the debate, drove the initiatives and profited from them. The existing utilities, interestingly, on the whole, were initially reluctant or even actively hostile. They threw their considerable political clout behind the issue only after the states agreed to force ratepayers to pay them tens of billions of dollars, in what can only in retrospect be viewed as bribes, to buy their support.

A serious debate about the kind of power rules needed for the 21st century is definitely in order. But the terms of that debate should not be defined by a few large suppliers and a few large producers and centered on customer choice of suppliers. Rather, we need a full-blown debate about the kind of electricity system we want.

One hundred years ago, at the dawn of the 20th century, we had such a debate. It lasted a generation and resulted in a hybrid electricity system one-third owned by customers and two-thirds owned by investors. The system was composed of local and regional electricity companies that owned both generation and transmission lines. In return for a guaranteed monopoly on the right to generate and sell electricity, these companies accepted an "obligation to serve," and agreed to be regulated by state and federal agencies whose decisions were to be, by law, "in the public interest."

That debate over the shape and scale of our electricity system occurred in a different historical context. Electricity was a new form of energy, and a rapid increase in demand was precisely what regulators and producers wanted. The quality of electricity didn't matter since the first applications required brute force rather than high quality electricity—lighting, heating and industrial motors. Environmental impact was not a consideration. Fossil fuels were presumed to last forever. No renewable fueled electricity alternatives promised low cost electricity in large quantities. And the technological dynamic favored ever-larger power plants and ever-longer transmission lines.

Today, in the immortal words of French writer Paul Valery, "the future isn't what it used to be." Policymakers now realize that we cannot continue to double electricity demand every ten years. Sensitive electronic equipment now demands a very high quality of electricity. Environmental considerations are paramount. Renewable electricity resources offer substantial amounts of power at increasingly competitive prices. And the new technological dynamic argues for smaller, more dispersed power plants and a diminishing reliance on long transmission systems.

This new reality demands a new way of thinking. Unfortunately, since Congress directed the Federal Energy Regulatory Commission (FERC) to change the wholesale power rules in 1992, policymakers have acted as if the future is exactly what it used to be. They have approved mergers affecting more than half the total assets of investor-owned utilities. Some predict with approval that by 2010 fewer than 10 giant electricity companies, many with a primary economic interest half a globe away, will provide the vast majority of our electricity. Authority has moved steadily away from the local and state level to Washington. And Washington's highest priority now is to increase long distance transmission capacity.

The electricity crisis in California, and the emergence of a new breed of electricity companies that sell small power plants, may serve to change the terms of the debate. Policymakers and customers are looking to regain control over their electricity system, bringing power, both figuratively and literally, to the people. In September 2000, California Governor Gray Davis signed a bill "to increase self-sufficiency of consumers of electricity." In his state of the state address, Governor Davis declared that all state universities and community colleges should "move toward energy independence." California Senate President pro tem John Burton proposed that the state purchase 32,300 miles worth of transmision lines from three private utilities. Why? "What we're trying to do here is give the state some influence and control over its own destiny."

The new technological dynamic argues for smaller, more dispersed power plants and a diminishing reliance on long transmission systems.

We are living through a historical moment in which renewable fueled electricity is increasingly competitive with fossil fuel generated electricity, and decentralized power plants are increasingly cheaper than central power plants. Even when these are more expensive, the increase in price is trivial. Back in 1993, California established an auction for clean power. It approved over 1200 megawatts of power contracts with wind, geothermal, cogeneration and other clean electricity suppliers. California utilities went to FERC and argued that California did not have the authority to do this because it would force the ratepayers to pay a few percent more for their power. FERC agreed, and in 1995 ordered

California to cease and desist. Five years later California has experienced rate increases of 10-20 percent, and rate increases of 30-40 percent are not out of realm of possibility. Penny wise and pound foolish is the description of the utilities and the federal government's actions here.

Back in 1982, I wrote a book called *Be Your Own Power Company*. It predicted the new decentralizing dynamic in electric generation and concluded with an observation I still believe is valid. "The economic attractiveness of decentralization is becoming ever more apparent. Yet to emphasize only the economic value of decentralization would be a mistake. The political and psychological value of a widely distributed capacity to produce a commodity as essential as electricity is equally important. Self-reliance was a major objective of the nation's founders. Benjamin Franklin once remarked, 'The man who trades independence for security usually deserves to end up with neither.'"

Over the last 100 years we relinquished control over our electricity production in return for a promise of lower prices and increased security. Today, at least for the citizens of California and Oregon and Washington and Idaho, that trade has resulted in both higher prices and less security.

We can do better. This book is intended to help us do just that. It describes the changing technological and political context for the new power debate, and wherever possible offers actual rules and strategies for policymakers.

This focus on rules is intentional. We make the rules and the rules make us. The rules we design channel entrepreneurial energy and investment capital and scientific genius in a certain direction. We need to change the power rules to channel that creativity into designing and building an electricity system more compatible with the needs and values of the 21st century.

For those interested in translating theory into practice, I urge you to come to the Institute for Local Self-Reliance's web site, www.newrules.org. The site has available both the best analyses of the new electricity system's birthing pains, and actual rules—regulations, ordinances, judicial rulings—that have been adopted or proposed by city councils, state legislatures and state and federal public utility commissions. Feel free to download any or all. Ask your local policymaker to introduce them.

And let the real debate begin.

David Morris
Minneapolis
February 2001

Acknowledgments

First, let me thank Daniel Kraker, who helped me monitor the often chaotic developments in electricity regulations, did his darndest to spruce up the writing and above all, prodded me to get it out the door. Even from a distance his presence was felt.

To those who willingly gave their time and intellectual energy to review drafts of this book, a special thank you—especially to Nancy Rader, Matt Patrick and Mark Glaess. Any errors that remain, either of interpretation or fact, are solely my responsibility of course.

My deep appreciation to Elizabeth Noll, a wonderfully no-nonsense editor who helped all the pieces come together, and to designer Holle Brian.

This book is part of the Institute for Local Self-Reliance's New Rules Project. Visitors to our website, www.newrules.org, will be able to keep abreast of new developments and review all of the regulations, statutes and policies discussed in this book, as well as many others that enable a democratic and sustainable electricity system.

To John Bailey, who has managed to master both web architecture and subject content to an extent I've rarely witnessed, my unreserved admiration. And thanks again to Daniel Kraker for making the website as informative and accessible as possible.

Twenty years ago this summer I completed a book entitled *Be Your Own Power Company*. It described the embryonic cracks in the electricity system that were resulting from a new technological and economic dynamic. That book was written at Blue Mountain Center in upstate New York. Much of the writing of this book too was done there, as always under the wise and generous nurturing of Harriet Barlow. A tip of the hat, both professionally and personally, to you, Ms. Barlow.

INTRODUCTION

Changing The Rules

We take electricity for granted. When we flip the light switch we expect the light to go on. We don't much care why or how that happens. And we don't care about the intricate workings of the complex generation and delivery system behind that everyday miracle, unless that system breaks down or rates skyrocket.

Which is why the recent electricity crisis—rolling blackouts in California, soaring electricity prices in Chicago, electricity system fluctuations that burden high technology users—has come as such a shock. The level of debate about the electricity system has reached an intensity not seen in a hundred years. Indeed, in some measure, the current debate about the future shape and structure of the electricity system at the turn of the 21st century is reminiscent of the great debates about same question that occurred throughout the country, and the world, at the turn of the 20th century.

At the turn of the century the nation was mesmerized by electricity's potential. People strove to master its intricacies. Great battles were waged over who would own and control the power plants and transmission lines. Would we have customer or investor ownership, local or absentee governance? Would we have monopolies or competition? Would electric companies be regulated by local, state or federal governments?

Today these same issues are re-emerging as the nation rewrites the rules that will determine the future shape, scale, and ownership structure of our $250 billion electricity system, the nation's third largest industry after health care and automobiles.

The changes are coming in rapid-fire fashion. At the national level, the Federal Energy Regulatory Commission (FERC) is writing the rules that will open the transmission lines to competition (i.e. wholesale competition), while twenty-four states (and the District of Columbia), with well over half the nation's population, have already introduced competition at the retail level.

Regrettably, the current debate has, until very recently, been far too narrowly focused. The central issue has been, "Should customers have the right to choose their suppliers?" It is a remarkably restricted definition of "choice." Moreover, the rush to judgment has come neither in response to popular demand nor as a result of clear evidence that the electrical system circa 1995 was broken.

Even the most fervent supports of customer choice (i.e. retail competition) concede the lack of any grassroots demand. "Citing surveys finding most consumers content with their electric service providers, [FERC Commissioner James] Hoecker called the public's general silence in terms of demanding customer choice 'positively deafening.'"[1] After the Texas-New Mexico Power utility withdrew its restructuring plan (titled "Customer Choice") when it encountered substantial public opposition, a utility spokesperson lamented: "We're trying to give our customers something that would be good for them, but this is apparently something they don't know they need."[2]

For states that acted early to bring retail competition to their electricity markets, the results are in. Customers—especially residential customers—are simply not choosing new suppliers. In Massachusetts, after nearly two and a half years of competition, less than one-tenth of one percent of households, representing less than two-tenths of one percent of the state's total electric load, had switched suppliers. Pennsylvania has done better. As many as 10 percent of Pennsylvania's resi-

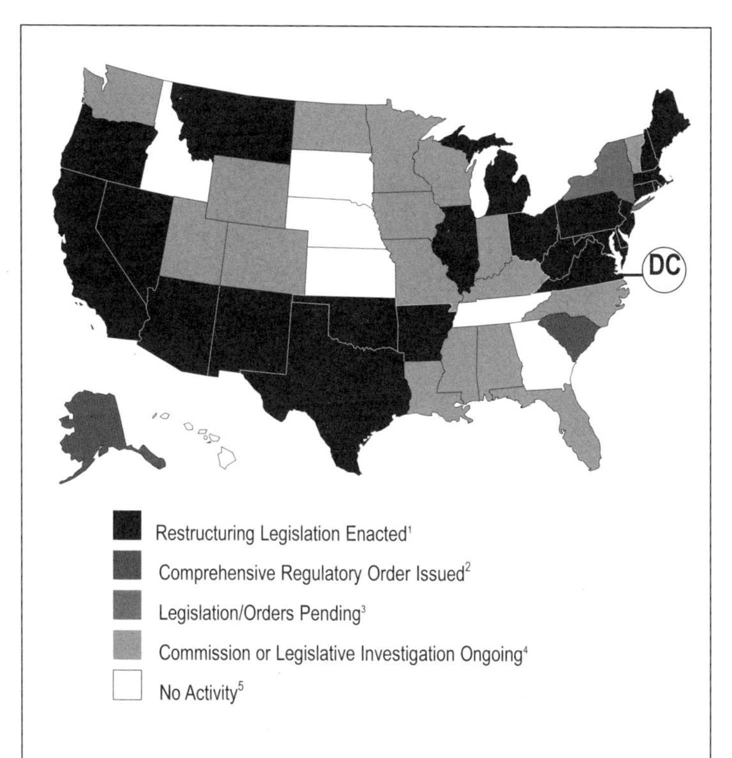

Fig. 1. Status of State Electric Utility Deregulation Activity, October 2000

1 Arizona, Arkansas, California, Connecticut, Delaware, District of Columbia, Illinois, Maine, Maryland, Massachusetts, Michigan, Montana, Nevada, New Hampshire, New Jersey, New Mexico, Ohio, Oklahoma, Oregon, Pennsylvania, Rhode Island, Texas, Virginia, and West Virginia.

2 New York.

3 Alaska and South Carolina.

4 Alabama, Colorado, Florida, Indiana, Iowa, Kentucky, Louisiana, Minnesota, Mississippi, Missouri, North Carolina, North Dakota, Utah, Vermont, Washington, Wisconsin, and Wyoming.

5 Georgia, Hawaii, Idaho, Kansas, Nebraska, South Dakota, and Tennessee

Source: Energy Information Administration, http://www.eia.doe.gov/cneaf/electricity/chg_str/regmap..html

dential customers have switched suppliers. But that is largely because the state's regulatory agency designed a rate structure that in effect penalizes customers who remain with their regular utility.

Fig. 2A. Percent of Customers Switched to Competitive Supplier (August 2000)[3]

State	Residential	Commercial	Industrial
California	1.7%	3.0%	13.2%
Massachusetts	0.09%	1.99%	11.8%
Pennsylvania	8.72%	16.07%	28.76%

Fig. 2B. Percent of Load Switched to Competitive Supplier

State	Residential	Commercial	Industrial
California	2.1%	10.9%	27.5%
Massachusetts	0.17%	3.95%	20.78%
Pennsylvania	8.52%	38.71%	48.53%

Source: California Public Utility Commission, Massachusetts Department of Environmental Regulation, Pennsylvania Office of Consumer Affairs.

Evidence from other monopoly sectors shows that, even when choice is offered, only a minority of customers participate. Fifteen years after long distance telephone was opened to competition, 54 percent of people still had not exercised choice. Two-thirds of all customers remained with AT&T.

A 1997 survey of all 50 state regulatory commissions by Martin Kushler identified only two—Maine and Vermont—that had conducted a scientific survey of utility customers to determine their opinions regarding utility restructuring.[4] Deregulation was not in and of itself a high priority. When residents of Maine were asked to choose between having utilities "deregulated to allow greater competition and possibly lower rates" or to "continue to be closely regulated in an effort to protect consumers and the environment," 54 percent preferred the latter.

People did express preferences, but for the kind of electricity choices that haven't been offered. Many, for example, preferred a more localized electricity system. Maine participants were asked, "Would you like to be able to choose your electric power provider if it meant the possibility of losing Maine-based utility companies to New England-based and nationally-based companies?" Fifty-six percent said no. In Michigan, a state which did a poll but not a scientific survey, the option of "buying additional power from another state or Canada" ranked dead last, at just 19 percent support, even lower than "building a nuclear power plant" (which had 21 percent support)!

Authority, Responsibility, Capacity—The ARC of Electricity

Electricity deregulation has occurred not because of popular demand or necessity, but because of two powerful players: independent power producers and large industrial customers. As *The National Journal* notes, "At both the state and federal levels, homeowners and small-business owners have been relegated to the sidelines in the electricity deregulation debate. The playing field has been dominated by business leaders who crave cheaper power and by new energy companies eager to serve the most lucrative customers."[5] The

Electricity deregulation has occurred not because of popular demand or necessity, but because of two powerful players: independent power producers and large industrial customers.

majority of electricity customers have become aware of the decisions made on their behalf only after the fact.

For proponents of deregulation, the only end is to allow customers to choose their electricity supplier. The only means is a virtually automated national marketplace for electricity.

But this is far too narrow a policy framework. The end—customer choice—defines us solely as consumers, ignoring our right and desire to choose systems that enhance our roles as producers and citizens.

The means—an unfettered marketplace—ignores the inability of the market to take into account important social values (e.g. universal service) or long-term costs (e.g. environmental degradation).

Everyone agrees that any future electrical system should be at least as safe, reliable, inexpensive and universal as the present one. Yet those standards can be achieved through a variety of ownership structures, generation technologies and fuel sources. We need to decide upfront about where we want to end up. The rules we create will channel entrepreneurial energies and scientific genius and investment capital in specific directions. What direction do we prefer?

This report proposes that policymakers develop rules that, as much as possible, shorten the distance between actors and those acted upon, between those who make the decisions and those who feel the impact of those decisions. This means decentralizing productive capacity, devolving authority, and devolving responsibility.

1. Decentralize capacity. The new power rules should encourage us to become "prosumers," to borrow Alvin Toffler's felicitous 25-year-old term. As Thomas Jefferson observed, the bedrock of a healthy democracy is the widest distribution of property. He defined property as wealth-producing assets. Technological advances offer us the opportunity to make decentralized capacity a key feature in our future electricity system, a literal way of giving power to the people.

2. Devolve authority. Policymakers should strive to decentralize not only electric power but political power by encouraging customer ownership of the distribution and transmission lines. By definition, customer-owned utilities are more democratic, located closer to customer-citizens and therefore more responsive to their values. The new power rules should not only nurture the capacity for self-reliance but for citizenship.

3. Accept responsibility. Electricity production is a key cause of environmental pollution, from acid rain to toxic wastes to global warming. The new power rules should force us to take responsibility for the impact of our consumption habits on future generations and on those who live outside our communities.

Decentralize capacity. Devolve authority. Accept responsibility. These objectives should guide policymakers at the local, state and federal level.

For over a century, we have been designing policies that move us in the opposite direction. Toward larger power plants and longer distances between producer and consumer and between those who make the decisions and those who feel the impact of those decisions.

The current debate about the future of electricity affords us an unprecedented opportunity to marry our kilowatt-hours to our values.

The crisis in California has made us rethink our approach to deregulation. Most states are now taking a wait-and-see attitude. But even in the states that have embraced retail competition, and at the federal level, where wholesale competition is the goal, the rules are not yet set in concrete. There is a great deal of fluidity in the decisionmaking.

There is much that can still be done to move us toward our goals even in those states that are further along toward deregulation. And in those states that have not yet embraced retail competition, the existing legislative and regulatory environments allow us significant authority, if only we will embrace it, to rewrite the rules to channel creativity and initiative in directions that embrace our three principles of authority, responsibility and capacity.

Seeing the Light is intended to provide a framework—an approach to the process of making decisions about our electric power system—and to help us learn from the mistakes and achievements of the industry's pioneers.

We cannot design the future without understanding the past. Thus this report begins by looking backward and discussing the interaction of policy and technology and structure in electricity's first hundred years.

The succeeding chapters discuss the present and future and the rules that can fashion a more democratic, environmentally benign and equitable electricity system.

CHAPTER ONE

A Whirlwind Tour of Electric History

1880-1900: Local Power

In the beginning, before there were giant utilities, state and federal regulatory agencies, regional transmission lines and rolling blackouts, companies didn't sell electricity. They sold power plants.

By the spring of 1883, the Edison Electric Illuminating Company had installed 334 generators in cotton mills, grain elevators, manufacturing plants, newspapers and theaters.[1]

When central power plants did emerge, they were often neighborhood affairs. One of the most widely publicized of the new central power stations was Edison's Pearl Street Station in lower Manhattan, which started up on September 4, 1882, and served 59 customers with a 72 kilowatt[2] load, with 1300 lamps over 12 city blocks. By 1890, over 1,000 central power plants were selling electricity to customers.

Competition among electricity providers was fierce. Cities offered electric companies the right to use the public thoroughfares to install poles and run wires, in return for compensation and local oversight. Cities rarely offered electric companies exclusive franchises. For example, in 1880 the Denver Common Council granted a city electric franchise "to all comers" with the sole restriction that "said companies do not obstruct the public thoroughfares."

New York City awarded six franchises on a single day in 1887. Chicago had more than 29 electric companies operating as late as the early part of the twentieth century.[3] The courts consistently ruled that in the absence of state legislative authorization, municipal corporations could not grant exclusive franchises for the ownership and operation of public utilities.

The Technological Dynamic: Bigger is Better

Technological advances quickly encouraged a more centralized system. Bigger plants could be also be operated at higher pressures and temperatures, making them more efficient.[4] In 1901 the Hartford Electric Authority installed a 2,000 kW power plant; in 1903 Chicago Electric Authority brought a 5,000 kW generator on line. Eighteen months later the largest plant was already twice that size. By the dawn of World War I the largest power plant could provide 35,000 kW, and by the mid 1920s, 175,000 kW.

The introduction of alternating current (AC) generators allowed high-voltage transmission lines to carry electricity economically from large, remote, inexpensive power plants to distant customers. In 1896 George Westinghouse opened a Niagara Falls power plant, which supplied the Buffalo Street railway 22 miles away. In 1902 a power plant in the San Francisco area transmitted electricity 200 miles.

Finally, customers realized that there was an inherent efficiency in being interconnected to many other customers and power plants. To be electrically self-sufficient, a customer had to install not only one power plant, but also a second one for backup. Moreover, customers had to size the power plant to meet their peak use, not just their average use. In the early years electricity was primarily used for lighting. Households and even businesses used lighting only for a few hours a day. The rest of the time the power plant stood idle. Also, households used lighting at different times—some people worked later, some left on weekends. Thus a power plant that needed to serve many households could be smaller than the total number of on-site power plants needed to serve individual houses.

The economics of connecting with the grid, plus the advances in generation and transmission technology, changed the structure of the electric industry. In 1900, 60 percent of electricity was generated on-site,[5] but as early as 1908 one observer noted that "although isolated plants are still numerous in Chicago, they were never so hard pressed by central station service as now..."[6] Firms began to abandon their power plants, slowly at first, and then with increasing speed.[7] From 1919 to 1927 some 52,000 small steam engines and an additional 18,000 internal combustion engines were scrapped. By 1930 only 20 percent of electricity was generated on-site.

Electricity is a unique commodity. Transporting it is not like transporting bushels of wheat or tons of coal or even information.

A distribution and transmission system is very difficult to manage. Electricity is a unique commodity. Transporting it is not like transporting bushels of wheat or tons of coal or even information. A grid system operator must not only prevent the distribution system from going down—cutting off electricity to customers—but from degrading the quality of the electricity by varying its frequency or voltage. This requires close coordination between power plants and distribution operators.

Early on it was accepted that the best way to sustain the reliability and quality of electrical systems was to have the ownership of the power plants and the distribution and transmission lines in the same hands. Bigger power plants were cheaper than smaller power plants. Interconnection was cheaper than going it alone. Duplicating distribution and transmission lines was a waste of money. Tight coordination between generator, transmitter, distributor and seller was essential.

By the early years of the 20th century, the vast majority of observers viewed electric utilities as natural monopolies, a situation in which, according to utility economist and former federal regulator Alfred Kahn, "as more output is concentrated in a single supplier...unit costs will decline."[8]

Who Will Control the Electricity Monopolies?

The complexity of transmitting electricity from multiple power plants to hundreds of thousand of customers argued for centralized control and ownership of the distribution and transmission lines, and many believed, of the generation plants as well. From 1900 on the principal question occupying policymakers was not "will we have competition?" but rather, "who will control the electric monopolies?" Would they be owned by customers or by investors? Would they be regulated by the communities which they served, or by distant state and federal regulatory agencies?

Customer-owned and investor-owned power systems battled one another for supremacy. As on-site power plants became central power plants, many cities began to establish municipal electric companies either because, as in the case of small cities, investors ignored their small markets, or, in the case of big cities, because investor-owned utilities often offered high cost, unreliable and unresponsive service.

By 1896, about 400 municipally owned electric plants were operating. By 1906 there were 1,250. Between 1902 and 1907 the number of municipally owned plants expanded more than twice as fast as privately owned plants.[9] More than 80 percent were in cities with fewer than 5,000 people, but several larger cities like Los Angeles, Seattle and Cleveland also opted for public power.

The financial crisis of 1907 and the default by New York City on its municipal bonds frightened investors and made it harder for cities to raise the capital to build their own electric systems. But even when cities did not directly own their power plants and distribution lines, they still retained the authority to issue franchises. Franchise agreements allowed a privately held utility to sell to city residents and businesses and use city streets to run its wires. Nationwide, a municipal franchise movement arose to provide technical assistance to cities who wanted to write the new power rules.

In 1910 and 1911, Delos Wilcox published his two volume work, *Municipal Franchise,* an exhaustive survey of existing franchise agreements throughout the nation. Wilcox urged municipal officials "to kindle a fire under every sleepy citizen till even the street gamins, the club women, and the great merchants on Broadway know what a franchise signifies."[10]

The municipal franchise movement was itself part of a broader home rule movement. Communities demanded more authority over their own affairs. The reaction by investor-owned utility companies was swift. As historian Leonard Hyman points out, "The idea [of state regulation] became increasingly appealing as a movement grew to make the electric utility business municipally owned."[11]

By the early years of the 20th century, the vast majority of observers viewed electric utilities as natural monopolies.

As electricity transmission spilled over city boundaries, investor-owned utilities argued that they were the superior organizational form. The investor-owned utilities trade association, the National Electric Light Association (later renamed the Edison Electric Institute), led by their President Samuel Insull, formerly an employee of Edison and later president of the Chicago Electric Authority (now Commonwealth Edison), aggressively lobbied legislatures to establish state regulatory agencies as a tradeoff for private electric monopolies.[12]

Investor-owned utilities argued that moving regulation from the city to the state was efficient. Stiles P. Jones, a utility expert for the National Municipal League, responded, "Efficiency gained at the expense of citizenship is a dear purchase and democracy plus efficiency is not unattainable."[13]

The first state regulatory commission was established in Wisconsin in 1907. By 1910 six state commissions were operating. By 1920 the number had risen to 35.[14] Today regulatory commissions exist in every state. In 8 states, regulatory commissioners are elected.

By the late 1920s, the structure of the modern electric system was in place. Utilities, both private and public, would be granted exclusive franchises to serve specific geographical areas. Transmission and distribution lines as well as power plants would be owned by these utilities. In return for an exclusive franchise these private monopolies would be directly regulated by state agencies. Customer-owned monopolies, for the most part, would be exempt from state regulation since their operations would be subject to the direct influence of their customers.

To attract the substantial amounts of capital required to build a rapidly expanding electricity system, utility investors would be guaranteed a healthy return. In return for this guarantee, the utility was expected to keep prices as low as possible.

The Federal Government Steps In

As electricity transmission had spilled across city lines in the early 1900s, by the late 1920s it was spilling across state lines. By 1935, 20 percent of the nation's electricity crossed state lines. A new organizational form—the electric holding company—emerged and rapidly rose to dominance. As early as 1914, 85 corporations controlled 69 percent of the nation's total installed generating capacity. By 1929, twelve controlled 76 percent; three controlled 45 percent.[15]

The foremost architect of the holding company was Samuel Insull. In 1912 he formed the Middle West Utilities Holding Company. By 1916 it controlled 118 power systems in 9 states.[16] By 1929, Insull controlled 239 companies operating in 30 states and Canada. One historian describes the Insull empire:

> The Insull interests controlled 69 percent of the stock of Corporation Securities and 64 percent of the stock of Insull Utility Investments. Those two companies together

> owned 28 percent of the voting stock of Middle West Utilities. Middle West Utilities owned eight holding companies, five investment companies, two service companies, two securities companies, and 14 operating companies. It also owned 99 percent of the voting stock of National Electric Power. National, in turn, owned one holding company, one service company, one paper mill and two operating companies. It also owned 93 percent of the voting stock of National Public Service. National Public Service owned three building companies, three miscellaneous firms, and four operating utilities. It also owned 100 percent of the voting stock of Seaboard Public Service. Seaboard Public Service owned the voting stock of five utility operating companies and one ice company. The utilities, in turn, owned eighteen subsidiaries.[17]

It became increasingly impossible for state commissions to supervise these byzantine accounting empires where costs could be shuffled between regulated and unregulated companies to maximize profits.[18] Will Rogers summed up the untenable situation: "A Holding Company is a thing where you hand an accomplice the goods while the policeman searches you."[19]

By the Depression the financial shenanigans and political power of the electric holding companies had sparked a wave of indignation, both at them and at a regulatory system that seemed incapable of ending their pernicious practices. People like Gifford Pinchot, Theodore Roosevelt's chief forester and governor of Pennsylvania in the early 1920s, railed against the power trusts in apocalyptic terms, "Nothing like this gigantic monopoly has ever appeared in the history of the world. Nothing has been imagined before that remotely approaches it in the thoroughgoing, intimate, unceasing control it may exercise over the daily life of every human being within the web of its wires."[20] In 1932, 37 congressional leaders from both parties signed a statement that declared, "The combined utility and banking interests, headed by the Power Trust, have the most powerful and widely organized political machinery ever known in our history."

The collapse of the stock market and the following economic contraction brought down the holding companies. More than 90 electric and gas companies fell into receivership. When the Insull enterprise crumbled, losses for investors in the company were estimated to be as much as $3 billion (equal to about $36 billion in 1998 dollars).

In 1932, 37 congressional leaders from both parties signed a statement that declared, "The combined utility and banking interests, headed by the Power Trust, have the most powerful and widely organized political machinery ever known in our history."

The federal government took a number of steps to ensure that the power trust would not be reincarnated. The Public Utility Holding Company Act of 1935 (PUHCA) established the Federal Power Commission (later renamed the Federal Energy Regulatory Commission) and charged it with the responsibility of encouraging "an abundant supply of electric energy throughout the United States with the greatest possible economy and with regard to the proper utilization and consideration of natural resources." The newly created Securities and Exchange Commission was given the authority to break up holding companies.[21]

Franklin Roosevelt wanted Congress to restrict utility holding companies to the business of operating and owning utility properties and to prohibit them from engaging in nonutility or speculative ventures.[22] This approach was defeated. Instead, utility holding companies were still allowed considerable authority to engage in nonutility activities. Even before the recent wave of deregulation, the nation's 150-odd utility holding companies owned over 4,200 nonutility subsidiaries. Regulatory oversight agencies have had to engage in the extremely difficult process of tracking internal transactions to prevent cross-subsidies and self-dealing from regulated to nonregulated utilities.

FDR also addressed the unwillingness of investor-owned utilities to offer electricity to sparsely populated rural areas. By executive order the President established the Rural Electrification Administration (now called Rural Utilities Services). REA provided long-term, low-interest loans to rural communities to build power plants and distribution and transmission lines. These new utilities were organized as cooperatives. For $5 rural residents could become members of a cooperative or a public utility district and collectively own their own power plant or distribution system, or bargain with previously recalcitrant investor-owned utilities for more modest electricity rates. Each customer had one vote.

The utility industry fought hard against PUHCA, but as several utility historians note, it seemed unconcerned with the establishment of the REA. One utility executive sniffed, "Let the farmers build electric cooperatives; then when they fail, we will buy them up at ten cents on the dollar."[23] The farmers shrugged and went out and built a reliable, low cost electricity system that served their needs. In 1933, only 11 percent of farms had electricity. By 1944, 43 percent of rural households had electricity. By 1975, 98 percent did.

In the 1930s, the federal government created several federal agencies to build and operate hydroelectric dams on America's most powerful rivers: the Tennessee Valley Authority, the Bonneville Power Authority, the Southeastern Power Authority, and the Western Area Power Authority. Between 1933 and 1941, 50 percent of all new power capacity built was provided by the federal government or other public power institutions.[24]

By law, these federal agencies were required to favor customer-owned utilities. This greatly benefited cooperatives, which often owned their own transmission lines.[25] But municipally owned power companies operating in areas of the country where investor-owned utilities (IOUs) owned the high-voltage transmission lines found themselves unable to benefit from this arrangement, because the IOUs refused to transmit electricity from federal dams to these cities. In the 1920s, Congress and the federal regulatory agencies refused to grant cities authority to establish joint authorities to build transmission lines.

When Congress debated the Federal Power Act of 1935 it considered whether the law should make the transmission lines common carriers, like roads, obligating power companies to transmit electricity for any person upon reasonable request. Such a provision was originally included in the House and Senate versions of the bill but ultimately Congress refused to adopt those provisions.[26]

That refusal would plague customer-owned electric companies for the next 60 years.

By the end of World War II, the structure of the U.S. electrical system was set. It was to be a hybrid system, largely owned by private utilities regulated by state and federal agencies but with a significant portion of the electricity transmission system and a modest portion of the generation system owned by customers.

Bigger is Better and Nuclear is the Biggest of All

Regulators adopted a cost-plus ratemaking system that encouraged bigger power plants and more transmission lines. The greater the investment, the higher the profits. Regulatory commissioners rarely second-guessed utility investments.

This regulatory system worked well in an era when electricity demand increased like clockwork and the cost of borrowing stayed low. This was the case from 1950 to 1970. During that period, fuel prices fell and power plants became more efficient. The overall result was that by 1965, the average price of electricity had declined to the all-time low of 1.5 cents per kWh.[27]

Utility stocks became a favored nest egg for widows and pension funds. The return on utility stock, guaranteed by the regulatory commissions, averaged a healthy 10-11 percent from 1948 to 1965.

But the doubling of electricity demand every decade eventually imposed a heavy burden on the system, for each doubling was on an ever-larger base. When electricity demand doubled from 1960 to 1970, utilities had to add 165,000 MW in capacity, the equivalent of 1,000 165-MW power plants, a typical plant in 1960. In 1964 the Federal Power Commission predicted that electricity demand would double again by 1980 and double again by 1990 and again by the year 2000. The country would need to add the equivalent of about 15,000 165-MW power plants in 30 years.

The only way to meet this rapidly mushrooming demand, policymakers believed, was to build equally gigantic power plants. They maintained that this was not only convenient but cost-effective. "In the 1950s and 1960s, it appeared there were enormous economies of scale," recalls Alfred E. Kahn.[28]

Fig. 3. Typical Size of Steam Electric Turbines 1900 - 1980

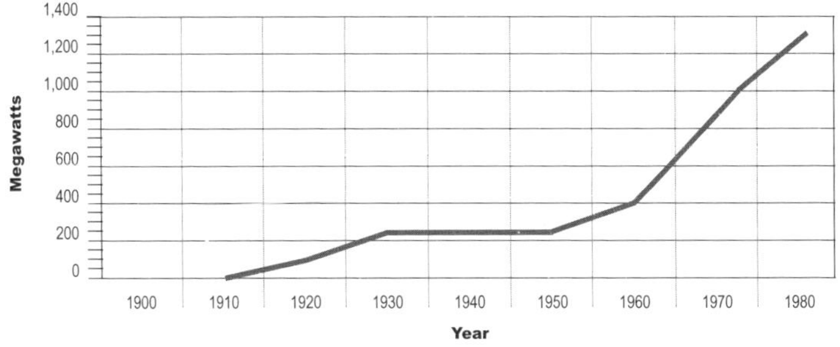

The size of power plants grew rapidly. In the 1920s, power plants that generated 30-70 MW (megawatts, equal to 1000 kilowatts), enough to meet the needs of about 50,000 households, were common. By 1945 the average size unit sold was still only about 40 MW.[29] As late as 1953, the 208 MW power plant that General Electric had put into operation in 1929 continued to be its largest unit.

But by 1967 the median-size unit ordered had soared to 700 MW. The largest units were over 850 MW.[30] By the end of the 1970s the size of the median power plant ordered climbed to 1000 MW. By 1977, of the 4,000 power plants in operation, fewer than 300 generated more than half the nation's power.

The need for bigger power plants to meet soaring demand coincided with the introduction of a new technology that encouraged the construction of the biggest power plants of all—nuclear-fueled generators. In 1954 the Atomic Energy Act (PL 83-703) allowed for the private development of nuclear power. In 1954 the first Chair of the Atomic Energy Commission, Admiral Lewis Strauss, infamously promised the nation that nuclear electricity would be "too cheap to meter."[31] The first commercial nuclear reactor opened in Shippington, Pennsylvania in 1957.

By 1972, the federal government made nuclear power the centerpiece of its electricity growth strategy. President Richard Nixon envisioned 1,000 nuclear plants operating by the year 2000, each with a capacity of more than 1,000 MW.[32]

Between 1971 and 1974, 131 new nuclear power plants were ordered, with an average capacity of 1,100 MW, in line with Nixon's goal. The White House Office of Science and Technology reported that "in the next twenty years, new capacity will come from 250 huge power plants in the range of 2000 to 3000 MW," large enough to supply 2-3 million households.[33]

The System Staggers

In October 1973, OPEC quadrupled the price of oil, and the shock wave staggered the electricity system. Inflation soared. So did the cost of borrowing. The price of new power plants rose steeply. The cost of nuclear power increased from $150 per kW in 1971 to over $600 per kW in 1976 and up to $1,200 per kW in the early 1980s.[34]

Ordinarily, utilities came to regulatory agencies on a regular level to ask for a rate change (for decades this meant a lowering, not a raising, of rates). But the rapidly rising fuel prices pushed utilities to the wall. In February of 1974 the New York State Public Service Commission granted Consolidated Edison a temporary rate increase of $75 million, but by April Con Ed's cash reserves were so low it was forced to omit a dividend payment on its common stock for the first time in its 89-year history. The price of utility stocks fell 38 percent that year.

In 1979, the near melt-down of one unit at Three Mile Island dashed the inflated hopes of nuclear power advocates. At the same time a second oil price hike destabilized the domestic and global economies.

To help keep utilities solvent, regulatory agencies allowed higher fuel prices to be passed through to customers automatically. In 1974 the real price of electricity rose for the first time since the depths of the 1930s depression.[35] In 1975 nearly $5.9 billion in fuel adjustments were passed on to consumers.

High interest rates raised the cost of power plants even while utilities raised more and more capital to expand the capacity to meet Americans' soaring appetite for electricity. Capital spending by U.S. electric utilities rose from $5 billion in 1965 to $15 billion in 1974 and to almost $30 billion by 1980.[36] As much as 40 percent of the consumer's bill was coming from financing charges.[37] Private power companies were issuing half of all common industrial stock every year and absorbing a third of all corporate financing. Half the income of major investment bankers was coming from financing private power companies.

In 1979, the near meltdown of one unit at Three Mile Island dashed the inflated hopes of nuclear power advocates.

At the same time a second oil price hike, even larger than the one in 1973, destabilized the domestic and global economies, generating shockwaves which to this day continue to ripple through the international economic and financial system.

Electric prices rose by 19 percent in 1980 and 15 percent in 1981. In 1981, regulators granted rate increases totaling $8.3 billion, 80 percent of what utilities requested. In 1982 the rate increase was $7.6 billion. The average price of electricity by the 3 major investor-owned utilities in California rose from less than 2 cents per kWh in 1965 to almost 8 cents per kWh in 1982.[38]

About 100 nuclear plants were canceled from 1972 to 1994, but utilities still fought to bring on line another 40 nuclear plants.[39] Those expensive nuclear plants form the principal battleground in the current debate about "stranded costs."[40]

Utilities had sunk $80 billion into nuclear plants and looked to invest another $45 billion by the end of 1987. Some of these utilities were building plants that cost more than 100 percent of their shareholders' equity. They were, in short, betting the company on gigantic new plants.[41]

And then, as the price of electricity rose, the demand for electricity slowed. By the late 1970s demand was doubling every 25-35 years, not every 10 years. In the early 1980s the nation went into recession. The demand for electricity actually fell for the first time since the early 1930s.

A wave of increasingly costly power plants were in the pipeline while customer demand growth was moderating. The regulatory system seemed incapable of stopping the impending train wreck.

In 1982, electricity demand dropped for the first time since 1931. Surplus electrical generating capacity reached 39 percent, more than double that recommended in the industry. In 1984 the Washington Public Power System defaulted on over $2 billion in bonds it had issued to build four nuclear reactors. In 1988 the Public Service Company of New Hampshire declared bankruptcy because of the cost overruns involved with its Seabrook nuclear power plant.

The underpinnings of the electrical system seemed to have come undone. A 1984 cover story by *Business Week* asked, "Are Utilities Obsolete?" Inside, the report observed, "The once-placid electric utility industry has never seen anything like it. As the money tied up in unfinished nuclear power plants has mounted to alarming levels, banks have turned skittish and investors have fled, raising the threat of bankruptcy for some... Most disconcerting of all, the time-honored system of supplying, pricing, regulating and financing electricity—a system that was not good enough to avert the present crisis—may be outmoded."[42]

"For six decades the task of state regulators was essentially one of distributing among rate payers the benefits of the progressively higher efficiencies achieved by utility managers," one commentator observed, "not bad work if you can get it, but you can't get it anymore."[43]

The State and Federal Response: Independent Power, Prudence Reviews and Full-cost Accounting

Congress and the states responded to the crisis of the 1970s and early 1980s by fundamentally changing the power rules. At the federal level, the 1978 Public Utilities Regulatory Policies Act (PURPA) abolished the 75-year-old monopoly utilities had held over the generation of electricity. Congress justified PURPA as a national security measure, a strategy to encourage the construction of renewable-fueled or high-efficiency power plants[44] that could wean us from our dependence on imported oil.

The 1978 Public Utilities Regulatory Policies Act (PURPA) abolished the 75-year-old monopoly utilities had held over the generation of electricity.

PURPA also required utilities to purchase power from independent producers at favorable prices and prohibited them from putting obstacles in the way of customers generating all or part of their own power.

Independent power

For some, no clear and present danger justified federal intervention into what were traditionally state affairs. In March 1981, the same month PURPA was to go into effect, Judge Harold Cox of the Southern District Court of Mississippi declared PURPA unconstitutional. Upholding the arguments of the Mississippi Power and Light Company, the state of Mississippi and the Mississippi Public Service Commission, he proclaimed, "The sovereign state of Mississippi is not a robot or lackey which may be shuttled back and forth to suit the whim and caprice of the federal government."[45]

In the spring of 1982, by one vote, the United States Supreme Court overruled Judge Cox.[46] The independent power industry was born. Applications to FERC soared from 30 in 1980 to over 500 by 1982.[47]

Governments at many levels offered independent power producers a helping hand. PURPA exempted them from most federal utility regulations. Congress provided handsome tax benefits not available to utilities.

When the price of oil skyrocketed in 1980, for example, New Hampshire doubled the price utilities had to pay for certain kinds of independent power from 4 cents to 8 cents. In 1982 Montana doubled its minimum from 3 cents to 6 cents. That same year, the New York State legislature mandated a minimum 6 cent per kWh rate. California established a 10-year "standard offer" contract that required a payment of about 10 cents a kWh, higher than the retail rate for electricity.

These prices seem very high, in retrospect (or at least before the recent runup of electricity prices in several parts of the country). But they were consistent with the projected prices of conventional new power plants.

Coincidental with the establishment of guaranteed high prices for independent power producers, the price of natural gas unexpectedly dropped. Since many of the independent power plants were fueled by natural gas, this gave these plants a major boost.[48]

Because they were not operating within a cost-plus regulatory system, independent power producers (IPPs) had an incentive to lower costs. They did so in a variety of ways.[49] One was by building smaller power plants that came on line quickly, significantly lowering capital costs. Competitive bidding for construction and fuel contracts encouraged lower prices by suppliers and technological innovations by equipment manufacturers. Independent power producers also learned to extract more useful work from a given amount of fuel. By the early 1990s the efficiency of their power plants was approaching 50 percent, a 40 percent improvement over conventional large scale utility power plants.

Fig. 4. Utility and Nonutility Net Additions to Nameplate Capacity 1986 - 1995

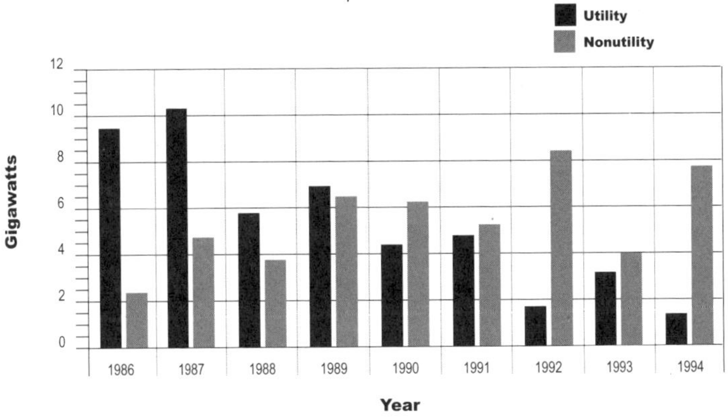

Source: *The Changing Structure of the Electric Power Industry: An Update*, Energy Information Administration, December 1996

From 1979 to 1992, 30 percent of all new electrical capacity added was built by independent power producers. By 1991, nonutility generators, some of whom were nonregulated utility subsidiaries, were adding more than half the nation's new electrical capacity. In 1994 they accounted for almost three-quarters of the new capacity added in that year.[50]

As their capacity has grown, so has their production. In 1992 nonregulated power producers were generating 7.1 percent of the nation's electricity. By 1998 this had risen to 12 percent.[51]

Prudence reviews

While independent power production flourished, state regulatory agencies took a more aggressive, hands-on approach. From the 1920s to the 1970s, regulatory commissions rarely excluded utility investments from the rate base. In the late 1970s and early 1980s, however, state regulatory commissions began to take an increasingly hands-on attitude toward utilities. From 1985 to 1991 regulatory commissions disallowed $14 billion in nuclear investments, forcing shareholders rather than customers to bear the loss.[52]

Utilities sued, arguing that since these plants had been approved by the commissions, any investment in the plant, even if unanticipated, should earn the guaranteed rate of return. The courts disagreed.[53]

Prudence reviews introduced an element of uncertainty about future profits for power plant owners. One unintended consequence was to encourage utilities to reduce their own power plant investments even further and purchase power from independent producers.[54]

Competitive bidding, least-cost planning, full-cost accounting

Stung by the cost overruns on nuclear plants, and urged on by the new independent power industry, regulators began to require that utilities undertake competitive bidding for new capacity. No longer would utility-owned plants be the only plant considered. Maine became the first state to adopt a competitive bidding requirement in 1984. From 1984 to 1991, 36 states adopted or considered competitive bidding procedures to acquire new capacity.[55]

Prodded by environmental activists, regulatory commissions also began to design rules that eventually became known as "least-cost planning" and later "integrated resource planning."

Prior to the advent of least-cost planning, a utility planned for and acquired new resources without the involvement of regulators or the public, except when it came to choosing sites for power plants. Least-cost planning required that utilities' resource acquisitions be scrutinized by regulators and the public in advance. Future demand growth was treated as an outcome of a planning process[56] rather than as a fixed input to that process. In other words, planners began to consider both supply and demand-side options. They began to examine whether it was cheaper to save a kilowatt than to build a new kilowatt of capacity.[57]

In the late 1970s, Zach Willey, Arjun Makhijani and Edgar Kahn, working with the Environmental Defense Fund, put meat on the theoretical bones of least-cost planning. They adapted a traditional utility model so that it took into account efficiency investments, and persuaded the California Public Utilities Commission that such a concept could be effectively used in utility rate proceedings.

By 1984, the National Association of Regulatory Utility Commissioners formally endorsed least-cost planning by creating a Committee on Energy Conservation.

That conservation of energy was often cheaper than new construction was rarely a controversial point. The problem, as Stephen Wiel, chair of the Nevada Public Service Commission, wrote as late as 1989, was that, "The current rate-setting process does not accommodate conservation." Conservation might be cheaper, but utility investments in power plants earned a profit. Utility investments in conservation did not. And conservation reduces sales and revenues, further discouraging utility participation.

For almost 20 years, regulators tried to fashion rules that would encourage power companies to get into the energy conservation business.

Some, like California and Massachusetts, allowed utilities to recover lost earnings. That prevented the utilities from losing money as a result of conservation investments but it did not enable them to make a profit. Others, like Oregon, allowed conservation investments to be included in the rate base and therefore to earn a profit. But this mechanism often encouraged utilities to invest in high-priced conservation schemes that saved little energy. Still others, like Wisconsin, allowed utilities to share in the energy savings of their customers.[58]

Utility investments in power plants earned a profit. Utility investments in conservation did not. And conservation reduces sales and revenues, further discouraging utility participation.

By 1988, 25 states had implemented or were in the process of implementing least-cost electricity plans; another 18 were formally considering or developing such plans.[59]

In 1990, the U.S. Energy Information Administration began formally tracking utility energy conservation expenditures, which grew from $900 million in 1989 to $2.7 billion in 1993.[60]

By the early 1990s, a number of states were beginning to add full-cost accounting to the notion of least-cost planning. That meant accounting for the environmental damage of power plants. For the first time state agencies began to quantify the environmental costs of power plants and use that cost when comparing power plant bids. By the early 1990s, half a dozen states had quantified the environmental cost of pollution from various power plants and were beginning to experiment with how to integrate these costs into the competitive bidding process. In 1993 California issued the first formal competitive bid restricted only to "clean" energy sources.

By the early 1990s, the shock to the electrical system had resulted in a profound change in the power rules and the way electric utilities did business.

CHAPTER TWO

The Crisis Ends: Restructuring Continues

The tidal wave of rising fuel prices, higher interest rates, slumping demand and excess capacity crashed onto the electric shore in 1979, and by 1982 had begun to recede. The system began to stabilize.

Power plant orders dropped. Increasing consumption shrank costly excess capacity. Fuel prices fell. So did the cost of borrowing. From 1982 to 1993 the average retail price of electricity declined by 26 percent.[1] The system appeared to have righted itself, and by 1995 the hybrid system of customer- and investor-owned utilities and local, state and federal regulatory oversight could look back at a century of development with understandable pride.

- the real price of electricity dropped by 98 percent;
- the efficiency of power plants improved fiftyfold; and
- the overall reliability of the electricity system remained very high.

Competition was increasingly characterizing the generation segment of the electricity system. Policymakers and regulators were learning how to reward utilities for investing in least-cost strategies that included improving efficiency rather than building new power plants. And through mandates, tax incentives and regulatory changes, policymakers had nurtured an embryonic but continually expanding renewable electricity industry.

Fig. 5. Real Retail Prices of Electricity Sold by Utilities 1960 - 1997

(cents per kWh, 1992$)

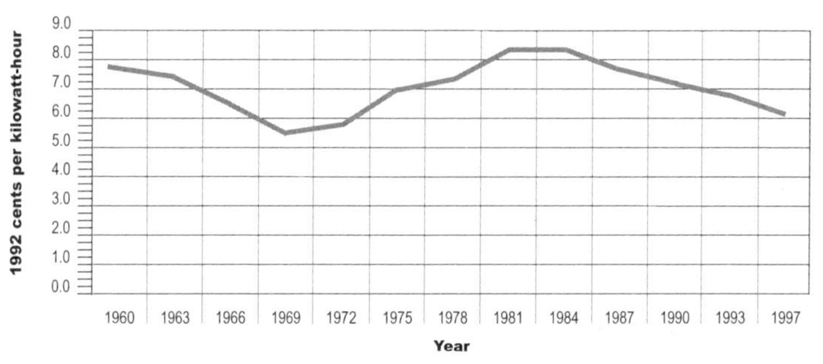

Source: Energy Information Administration, Annual Energy Review 1997, Table 8.13

Did Regulation Work?

The evidence is unclear. The price of electricity fell by more than 65 percent from 1900 to 1932. But some believe it would have fallen even more without regulation. In 1935, the Federal Power Commission (FPC) compared unregulated and regulated states and concluded, "there is a basis for the widespread feeling. . . that state regulation in general has failed to achieve its primary purpose." The FPC found that profits in regulated states were excessive. "[R]egulated companies are earning higher profits than companies which operate in states where regulation is delegated to municipalities."[2]

Thirty years later economists George Stigler and Claire Friedland re-examined the data from that era using more sophisticated statistical techniques. Their conclusion was the same as that of the FPC. Regulation had no impact on the return to shareholders nor the price to ratepayers.[3] In 1979 University of Rochester economist Gregg Jarrell concluded that prices and profits rose sharply in newly regulated states. Utilities appeared to benefit most from regulation and they promoted regulation in states where price wars were the most intense.

Unsurprisingly, perhaps, researchers found that in those states that had elected regulatory commissioners

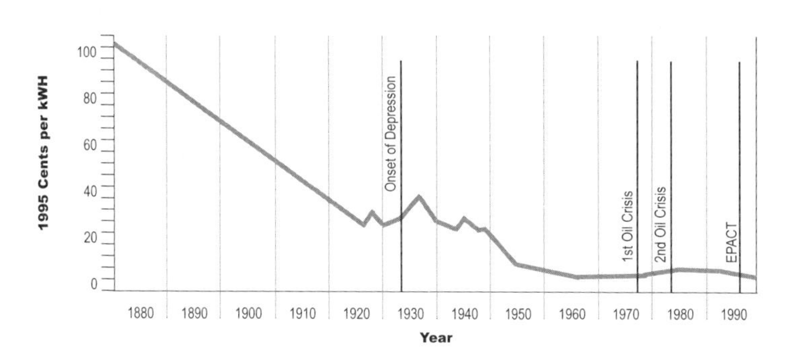

Source: Energy Information Administration

electricity prices were lower than in states whose regulators were appointed.[4]

Some historians believe that competition, not regulation, was the key to low prices and good service. Walter Primeaux tested this hypothesis by examining cities that had competing utilities.[5] Customers could switch back and forth between utilities. Some cities had duplicate distribution systems, a redundant investment that should have led to higher prices. But Primeaux found that in practice, costs to the utilities for providing electricity were lower, customer prices were one-third lower and the quality of service was much higher in cities where utilities engaged in direct competition.[6] The benefits of competition offset the advantages of larger scale and the increased investment in distribution lines.

Competition usually took the form, not of competition within a city, but competition between customer- owned and investor-owned utilities. "Approximately 4,000 cities, towns and villages have supplied electricity to their inhabitants at one time or another," writes Leonard Hyman. "But statistical measures of municipal electric enterprise fail to reveal its full significance because unsuccessful attempts to establish municipal systems leave no mark in the census reports, although they often influence the rates and policies of private companies threatened with condemnation or competition. Private companies which do not compete directly with municipal systems frequently find that their rates are compared with those of successful municipal systems or even the probable rates of proposed public systems."[7]

While the electricity system had stabilized, and state agencies had established rules that were guiding utilities toward a more environmentally benign and cost-effective future, the forces unleashed in the 1970s and '80s had dramatically changed the external context in which the debate over regulation was taking place.

At least four elements characterized this new electricity environment, which together set in motion the dramatic move by states to bring retail competition to the industry.

1. Powerful non-regulated independent power producers (IPPs)

By 1991 IPPs accounted for a majority of all new generating capacity added. By the mid 1990s IPPs were no longer small businesses. Six of the top eleven IPPs and ten of the top 20 were electric-utility affiliates.[8] Southern California Edison's subsidiary, Mission Energy, for example, was the largest "independent" in terms of ownership of nonutility generating projects.

As independent power producers grew in size and reach, they began to chafe under the existing regulatory and legal restrictions. They wanted to sell to a wider market. Teaming up with customer-owned utilities, they lobbied Congress to allow them equal access to utility-owned transmission lines to expand sales to more distant customers.

Teaming up with large customers, independent power producers also lobbied Congress and state legislatures and regulatory commissions to allow them to sell directly to electricity consumers, a concept known as "retail wheeling."

2. Small on-site power plants

By the early 1990s a century-long technological dynamic that favored ever-bigger power plants was coming to an end. In

1980, a 700-1,000 MW nuclear-fueled power plant cost as much as $1.5 billion. Two decades later a $50-75 million 100-MW gas-fired power plant can compete with many of its larger brethren. By the early 1990s IPPs could install a power plant in a large factory or even a large office building and that site could generate electricity at a lower price than the utility company was charging.

Big customers threatened to install their own power plants, leading utilities to ask regulatory commissions to allow them to negotiate individual contracts at lower prices with such customers. They argued that if their large customers left their systems, they would have to significantly raise the prices for the remaining customers. Most commissions acquiesced. This dramatically restricted the potential for on-site power plants, but in high-cost electric states the threat of a mass exodus by industrial customers remained a challenge to utility regulators.

3. A widening disparity of electricity prices

In 1970 electricity prices, on the whole, did not vary greatly between states. The national retail rate was about 1.7 cents per kWh while the variation among virtually all states was less than half a cent.

By 1990, the disparity in electricity prices among states had widened considerably. Those states that had bet heavily on nuclear power or that had forced their utilities to pay high prices for independent power found themselves saddled with significantly higher prices than their more cautious neighbors. The average electricity price was about seven cents and the variation among states could be more than five cents.[9] These high-priced electricity states became fertile ground for radical electricity reform. Industrial customers led the fight.

4. The changed political climate

In the 1970s a Democratic administration deregulated trucking and airlines. In the 1980s a Republican administration and the courts deregulated natural gas and telephones. In the 1990s a Democratic administration continued to embrace the logic of deregulation. Monopolies and government regulation were seen as a relic of another technological and political era.

Electricity came to be viewed as the next logical sector to feel the cleansing winds of competition and deregulation.

The combination of an increasingly powerful independent power industry, a wide disparity in electricity rates, the advent of smaller power plants and the national consensus in favor of competition proved irresistible. The dam burst in 1992 with the passage of the Energy Policy Act (EPAct). EPAct directed FERC to make transmission lines common carriers. All electricity suppliers, whether utility or nonutility, should have the same access to transmission lines and pay the same price for similar services.

The Energy Policy Act proved a watershed law, for it set in motion a dynamic difficult to stop. The intimate relationship between wholesale wheeling and retail wheeling and between federal and state regulatory authority made it hard to confine the dynamic simply to the wholesale market. Congress specifically prohibited FERC from mandating retail competition, but while FERC's jurisdiction is limited to sales for resale and interstate transmission, this still affects about 29 percent of all electricity sold or exchanged in the U.S.[10]

In 1993 another event occurred which may in retrospect be considered a watershed event in the history of electricity regulation. California's state agencies had concluded that a projected increase in electricity demand would require new power plants by the end of the decade. In order to deal with the issue of environmental costs, California's Public Utility Commission (CPUC) decided to hold the nation's first competitive bid for electricity restricted to "clean" electricity sources. These included wind power, geothermal, solar energy, biomass and high-efficiency natural gas-fired cogeneration systems. In 1993 the CPUC approved bids for 1200-1400 MW of capacity. The prices ranged from 3.68-3.85 cents per kWh for an on-site cogeneration power plant in the Hunters Point section of San Francisco to 5.73-6.49 cents per kWh for power generated from geothermal, wind and biomass.

California utilities appealed to the Federal Energy Regulatory Commission (FERC), arguing that they did not need new power and even if they did, California did not have the authority to require them to pay prices higher than they could buy power from the conventional wholesale market. Southern California Edison claimed that it had "more than enough power for the next ten years."[11] San Diego Gas and Electric claimed its ratepayers would be forced to pay premiums running into the hundreds of millions of dollars over the life of the proposed contracts. FERC agreed with the utilities. It voted 5-0 to ask the CPUC to halt the auction. The federal government had preempted California's right to restrict new electricity sources to those generated from clean fuels. In March 1995, the CPUC stayed its auction approvals.

The federal government had pre-empted California's right to restrict new electricity sources to those generated from clean fuels.

The preemption of California's authority reinforced and perhaps accelerated its embrace of a radical deregulation plan. A year after it suspended its clean energy initiative, California's legislature enacted legislation that dramatically deregulated California's electricity market. Two years later the deregulation was fully in effect. Two years after that, California was suffering sever price increases. In desperation, it established another auction, this time for all sources of electricity. The average price was over 7 cents per kWh. Meanwhile, California ratepayers were facing the prospect of having to pay more than ten billion dollars in increased rates.

California was the first in the nation to open its markets to retail competition in early 1998.[12] By early 1999 six more states had fully opened their markets to retail competition. By late 2000 that tally had grown to 24 states, plus the District of Columbia.

Four Arguments Against Retail Wheeling

Wheeling in electricity parlance means transporting. Wholesale wheeling means transporting across transmission lines from supplier to wholesaler, usually the utility, which then resells the electricity to retail customers. Retail wheeling means transporting across transmission and distribution lines directly to the final customer. In 1978, as discussed earlier, Congress allowed independent power producers to sell wholesale to utilities. But utilities were not required to transport (wheel) the electricity across their lines to distant utilities. And utilities often forced independent power providers to pay higher costs than the utility-owned power plants to use the utility-owned transmission system. Thus in 1992, as the following section will discuss in more detail, Congress required that utilities treat independent power suppliers the same as they treated their own power plants, giving them equal access at an equal price to transmission capacity. FERC has been working for the past eight years to develop the rules to enable that objective. States, except for Texas, whose electricity system is virtually unconnected to that of other states' systems and therefore is largely exempt from federal authority in this respect, do not have the authority to make the rules regarding wholesale wheeling. They do have the authority to make the rules regarding retail wheeling, and that's where their focus has been.

Although there is a near-consensus on the need to make the transmission system more accessible to nonutility owned electricity generators, many experts have serious reservations about the speed at which we are trying to move toward retail wheeling. Here are four of their most significant arguments.

The transition to wholesale wheeling must be perfected before we introduce retail wheeling. Although Congress mandated wholesale wheeling in 1992, the country is still working out the details. The process is remarkably complex, requiring not only new regulations but new governance structures and perhaps even new technologies. The price hikes not only in California but in Illinois and New York City in 2000 and the growing concern about the overall reliability of our interstate transmission systems under the new system of wholesale competition testify to the fact that we have much to learn. Retail wheeling cannot occur without wholesale wheeling, but wholesale wheeling can, and should occur, without being burdened by the pressures that result from moving toward retail wheeling at the same time.

Virtually all projected savings from deregulation will come from wholesale, not retail, competition.

By the mid 1990s the majority of the population lived in communities where their utilities were required to issue competitive bids for new capacity. The Energy Policy Act of 1992 converted the transmission system into a common carrier where the owners of the transmission lines could not discriminate against independent power suppliers and in favor of their own power plants when it came to access to the delivery system. Presuming sufficient transmission capacity (a presumption discussed in another section), low cost or underutilized power plants in one region will be able to send their electricity to other higher cost regions nearing capacity. The resulting savings comprise virtually all projected savings from deregulation. Indeed, the Energy Information Administration has estimated that the additional benefit from retail wheeling by the year 2010 would be in the 1 percent range.

Retail wheeling cannot occur without wholesale wheeling, but wholesale wheeling can, and should occur, without being burdened by the pressures that result from moving toward retail wheeling at the same time.

Fig. 7. Electricity Prices by Region in Reference and Competitive Generation Scenarios, 2010

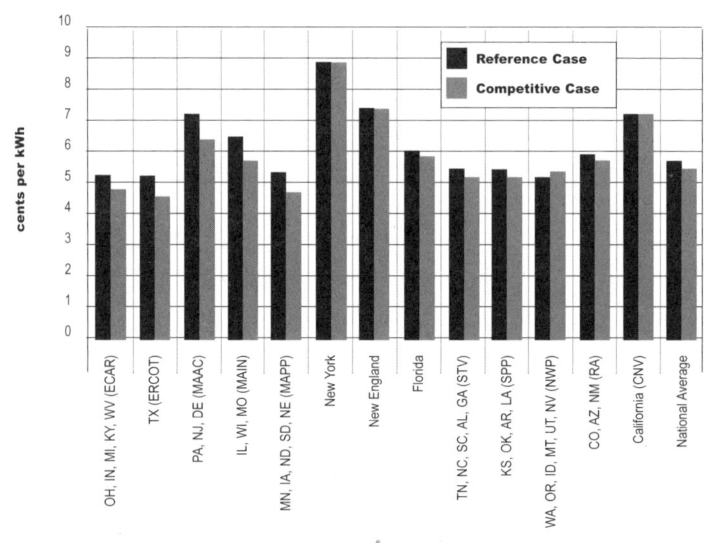

Source: Energy Information Administration, *Issues in Midterm Analysis and Forecasting 1998*, July 1998.

Wholesale wheeling allows any independent power producer to sell into the grid on an equal basis with utility-owned power plants. This is a logical next step after the implementation of PURPA in 1978. Wholesale wheeling allows virtually all of the objectives that people are demanding of their electricity system—decentralized power generation, renewable energy, energy efficiency, environmental protection. It is unclear whether retail wheeling encourages or inhibits the realization of these objectives.

Retail competition will generate modest if not trivial savings for most customers.

The Department of Energy estimates that if all Americans choose their electricity suppliers, as of January 2, 2003, the average electricity customer would save less than 10 percent on his or her bill. Big customers may save considerably more. Small customers will probably save considerably less. Glenn

Lovin, executive director of the Power Marketing Association, an industry trade group, concedes, "Many users have unrealistic expectations of savings. Even after stranded costs have been worked through, most commercial users will only see savings of 5 to 10 percent. . ." According to the Electric Power Research Institute, large commercial and industrial consumers in California that have switched suppliers are averaging savings of only three to five percent.[13] And that was before the huge rate increases that hit customers in early 2001.

In fact, rates for small customers may well rise under deregulation compared to what they would have been without it. While electricity prices rose in the 1970s and 1980s, by the mid 1990s they had returned in most states to near historic lows. Until natural gas prices began to soar in late 2000, the federal government projects continued annual price reductions of 0.5-1.2 percent for the foreseeable future.[14]

Deregulation vastly increases marketing overhead. Already several layers of intermediary buyers have emerged (aggregators, brokers, marketers). Customers must learn the art of negotiating electricity prices in the face of a bewildering variety of contract possibilities (e.g. firm and non-firm power, peak and off-peak rates, etc.). Rates may rise under deregulation because power plants are never paid off. Under the present system a power plant is amortized, which means it is gradually removed from the rate base. Under a deregulated system, capacity will never be removed from the rate base. Each new investor will demand an equally high or higher rate of return. And as of April 1998, the 50,000 MW or so of generating capacity that had been available for sale had been sold at about 1.7 times book value, virtually guaranteeing higher prices.[15]

Virtually all the goals endorsed in this book can be achieved without retail wheeling. States can, and have, required significant increases in the use of renewable energy without requiring retail wheeling. States can, and have, required significant increases in utility spending on energy conservation without requiring retail wheeling. States can encourage decentralized electricity production without requiring retail wheeling. On-site generators will be sized primarily to displace on-site electricity consumption. Any excess electricity generated can be sold to "the grid," that is, to the electricity marketplace at wholesale prices.

Wholesale Wheeling: Working out the Bugs

In the beginning utilities were largely stand-alone operations. They connected their customers to their own power plants, but had few connections to other utilities. Later, a network of high-voltage transmission lines spread out across the country and electricity moved between utilities. After 1950, when larger coal and nuclear plants were built, the voltages of transmission lines increased as well, from a maximum of 287 kV to the current maximum of 765 kV, in order to move electricity hundreds of miles between the power plant and the ultimate customer. In the 1990s utilities began to build long distance, high-voltage direct current (DC) lines, which could carry three times the amount of electricity as the conventional alternating current (AC) lines.

Initially transmission systems were constructed for reliability purposes. Just as customers found they could save money by relying on the grid system, so utilities found they could reduce the amount of investment in excess capacity and still meet reliability standards by developing stronger trading relationships with neighboring utilities. A study by one Kansas utility in the 1970s found that it could halve its reserve capacity by interconnecting with neighboring utilities and purchasing electricity on the wholesale market.

In the 1980s the transmission system, constructed for reliability purposes increasingly became the highway for energy transactions between utilities. Bulk power transactions grew six-fold from 1961 to 1987. In 1996 nearly 55 percent of all electricity consumed was purchased by utilities from other utilities and nonutilities.[16]

The Energy Policy Act of 1992, along with FERC's subsequent Orders 888 and 889, ushered in the new age of whole-

Fig. 8. Quantity of Electricity Traded by Quarter 1995 - 1997

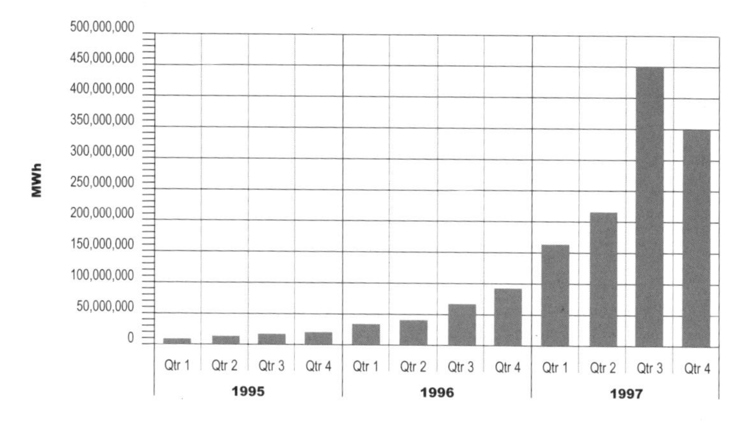

Source: Harris, Kiah E., P.E., *Life After 888 And 889*, Burns & McDonnell Engineering Company, July 1998. Prepared for American Public Power Association.

sale wheeling. Wholesale wheeling means the transmission of electricity from a supplier to a utility. This is different from retail wheeling in which the supplier sells directly to the final customer. We are still very much in the learning stage of that transition. Many problems must be solved.

Electricity is not like other commodities. Therefore, deregulating the transmission of electricity is not the same as deregulating trucking or airlines.

Electricity rarely travels along the contract path. The owner of a power plant in Montana might sell power to the owner of an office building in Seattle, but the electricity generated in Montana may well go to Los Angeles because of

The owner of a power plant in Montana might sell power to the owner of an office building in Seattle, but the electricity generated in Montana may well go to Los Angeles because of the physics of electric flows.

the physics of electric flows. One analysis concludes, "a power transfer from Indiana to New Jersey would produce flows over the lines of more than 20 different utilities and power pools. Less than half of the transferred power would flow over what would appear to be a relatively direct route."[17]

Thus managers of the transmission system can know how much capacity they may have on any one of their "highways" only when they know how much electricity is flowing on all of their roads.

As Kiah Harris points out, as large transfers of energy load up the transmission systems, utility systems along the way see voltage problems. Low voltages result from loop flows. To improve low voltages, utilities provide var support. "Unlike watts, vars cannot be transported long distances. They have to be produced locally where they are needed." Thus the local utility must compensate for low voltages by running power plants. Who does it charge? This issue has yet to be worked through.[18]

These so-called ancillary services of maintaining constant voltage and frequency levels and sufficient backup and reserve power, according to one utility expert, cost about $10-15 billion a year, about 50 percent of the cost of transmission. Yet as of mid 1998 Eric Hirst could observe, "there is still little consensus on how these essential transmission functions should be unbundled, produced, managed, delivered, measured and priced."[19]

Managing the electricity system has been made even harder because of the rise of a new entity in the electricity market: electricity marketers.

Traditionally, the relationship among buyers and sellers of electricity was almost familial. Since utilities owned both the transmission lines and the power plants and since they had a monopoly on sales within their own service area, they freely shared economic and technical information with one another.

In a traditional exchange, a system operator would contact operators of surrounding systems an hour ahead of delivery. When power could be purchased more cheaply from a power plant owned by utility B than by utility A, that electricity would be purchased by utility A and the two companies would split the savings.

The new electricity market is far less collegial. As FERC began to open up the transmission system, a new electricity player joined the team: the power marketer. On May 19, 1986 FERC approved Citizens Energy Corporation's application to become the nation's first power marketer. Only three more marketers were authorized by 1990. By the end of 1993 that had grown to 11, by early 1995 to 60, by early 1997 to 284 and by October, 2000 to over 450. These marketers, notes one observer, "trade it all—gas, electric and other commodities—they have a highly sophisticated integration of financial products, including options, futures, forwards, collars, hedges and other instruments to back their activities in the physical markets."[20] The largest of the power marketing firms now sell more power than some utilities. Enron is now the country's largest utility.[21]

In 1994 power marketer sales comprised about 1 percent of wholesale sales. By 1996 this had grown to 20 percent.[22] "Watching electric companies deal with one another used to be about as exciting as watching cows graze," writes Allan Sloan in the *Washington Post*. "The herd members were ultra-polite. They traded power back and forth, but no one gouged, because the guy you gouged today might be in a position to gouge you tomorrow. But since deregulation began in the electric biz a few years ago, a whole new bestiary has emerged. Bye bye cows. Hello, independent electricity traders; sharp-toothed velociraptors willing to bite, slash and maim to make a buck."[23]

The *Wall Street Journal* describes the new era of electricity dealing this way: "In this new world, wholesale electricity deals are made for one of two reasons. There is still the traditional need by utilities to find extra power or to sell it. But deals are also now being done simply to profit from trading—one trader will buy electricity from another in hope of being able to resell it quickly to a third trader at a profit."[24]

On average, the power is sold seven or eight times before it is delivered. Indeed, the market in trading electricity has already dwarfed the market for electricity itself, reaching the $10 billion level in 1997.

Jim Lee, vice president of The Power Company of America, an energy marketing firm based in Greenwich, Connecticut, predicts the annual volume of wholesale electricity commodity trading may reach $2.5 trillion by the year 2003. This is roughly 10 times the retail value of electricity trade in the US for year 2003.[25]

In the newly deregulated market, notes one observer, "[A]lmost anyone can become an electricity marketer, without assets, a reputation or a credit check," and "no one guarantees delivery if a player defaults."

This combination is a volatile one. Consider what happened in June 1998. A heat wave in the Midwest boosted demand while outages in several nuclear power plants reduced supply and storms knocked out several transmission lines. A power marketer called Federal Energy Sales defaulted on its obligation to provide power to the municipal utility of Springfield, Illinois, and to another trader, the Power Company of America of Greenwich, Connecticut. This failure led to defaults by Springfield and the Power Company, both of which had resold the power. A chain reaction set in as utilities feverishly searched for electricity to meet their customers' demands and paid any price to get it.

The short term price for electricity soared a thousandfold. Utilities serving Ohio and Illinois lost tens of millions of dollars by paying $5 for power they could sell to residential customers at only a dime or so. Illinois Power, for example, had expected replacement power costs in 1998 to be about the same as a year earlier, about $36 million. But with the price hikes it expected these costs to reach more than $130 million.[26]

When the dust had cleared, LG&E Energy Marketing, one of the first utility affiliates to enter the energy marketing business in 1984 and the fourth biggest marketer in 1997, had taken a huge loss and decided to abandon the marketing business.[27] John McCall, LG&E's Executive Vice President called the current electricity market "immature and not behaving rationally" and commented, "I don't think anyone anticipated the kind of ups and downs the market saw. No one anticipated prices moving as high as they moved."[28]

LG&E Energy Marketing sued City Water, Light and Power of Springfield for $21 million. El Paso Energy and Southern Energy Marketing also sued Springfield for $7.5 million. "We're just one piece of a daisy chain in all of this," moaned Bob Rogers, corporation counsel for Springfield. Springfield announced that it too would stop marketing power from third parties.

Then only one month later, on July 13th, prices in California over a five-hour period climbed to $9,999 per mWh, the highest possible price since California's new state trading entity's bidding software only accommodates four digits. The wholesale price rose to 10 cents per kWh, a 300 percent increase over the 2.5 cents during the entire year before the market was deregulated.

Were these two price spikes what life would be like in a deregulated market? Some power marketing companies argued that they were simply signs that restructuring had to be accelerated, that a fully open market would operate more efficiently and avoid such problems. However, Moody's Investment Service warned investors that the failures of two major power brokers to deliver power, "are not likely to be anomalies, and...additional failures by power marketers should be expected given the perhaps unsustainably large number of players in this nascent market and the fragile financial positions of several of these companies."

Southern California Edison was concerned that California's price spike, "unlike the recent problems in the Midwest, were not the result of extraordinary operating conditions...It was just an early summer day with near normal temperatures and no unusual outages." SCE observed that two companies—Houston Industries and AES—at the time controlled 1,000 MW and 700 MW respectively, sufficient to set the market clearing price.[29] A company might "game" the system by holding off delivering electricity at a crucial time, driving the price up. In California, such gaming appears to be occurring. According to the Energy Institute at the University of California at Berkeley, between June 1998 and November 1999, California consumers were charged $800 million above the prices that would have been charged in a truly competitive market.[30]

Several utilities and industrial customers petitioned FERC to step in and impose a cap on price hikes the way price caps have been imposed on other commodities.[31] FERC declined.

Today the transmission system is being stretched to its limits in many places. Some worry that this could degrade the reliability and quality of electricity. In the past utilities operated transmission and generation capacity conservatively, according to preventive operating procedures, to guard against system disturbances. These procedures included having more generators running than one would need to in normal circumstances, and limiting power transfers to leave transmission capacity available for emergencies. Today much of that excess is being sold on the market. *Newsweek* discovered that earlier in 1998, when supplies seemed ample, utilities sold power to outsiders to generate additional revenue. Were it not for those outside commitments, ComEd and Illinova and First Energy might have escaped unscathed.

According to the Energy Institute, between June 1998 and November 1999, California consumers were charged $800 million above the prices that would have been charged in a truly competitive market.

Basin Electric Power Cooperative Chief Bob McPhail worries about the impact of FERC's new rules. "Before this great government experiment started, the United States had the most efficient and reliable power delivery system in the entire world…There's a real possibly that electricity could become an unreliable product…"[32] The Philadelphia Electricity Company (PECO) concurred, telling FERC, "What used to be at times a severe, localized problem for certain transmission providers—parallel path flows—has now been turned into a severe, regionalized problem, which in the worst of times, may impact almost half the United States."[33]

The North American Electric Reliability Council recently concluded, "Although uncertainties and assumptions have always been part of long-term transmission studies, the level of uncertainty has increased tremendously. Purchases from undisclosed resources and the reluctance of generation developers to disclose plans for future capacity additions are making modeling for long-term transmission analysis virtually impossible."[34]

To some in the industry, the many problems involved in moving a transmission system that almost inherently depends on freely shared information and centralized control to one that relies on an autonomous market where the participants are not compelled to share information and indeed, have great incentives to withhold information to extract higher prices whenever possible, will require a great deal of patience, trial and error and learning. Kiah Harris, a utility expert with Burns & McDonnell Engineering Company, argues "The wholesale process has to be working before retail choice has any chance of success."[35]

And we are only just beginning to learn how to manage that process. In New England, for example, the region's first spot market for wholesale power wasn't established until December 1, 1998. The market sets an hourly price for short-term electricity, much like the stock market establishes a stock price.

Those states that have not yet embraced retail wheeling might be counseled to proceed cautiously. Learn from the pioneers. Refine the existing rules where possible. If dramatic cost reductions will not likely occur, it is all the more important to fashion rules that achieve objectives supported by the vast majority of customers.

It is to these objectives, and these rules, that we now turn our attention. In the following chapters we discuss existing proposals, and develop a few of our own, that will democratize, localize and downsize the electric system—and make it less burdensome on future generations.

CHAPTER THREE

Decentralizing Capacity: Bringing Power to the People

The recent growth in popularity of distributed generation is analogous to the historical evolution of computer systems. Whereas we once relied solely on mainframe computers with outlying workstations that had no processing power of their own, we now rely primarily on a small number of powerful servers networked with a larger number of desktop personal computers, all of which help to meet the information processing demands of the end users.

—National Renewable Energy Laboratory

- In 1999 Allied Signal introduced a 75 kW, $45,000 washing machine-sized power plant capable of supplying the electrical needs of a small store or restaurant.
- In December 1998, Capstone Turbine Corporation shipped its first two 30 kW microturbines. In November 2000 it shipped its 1000th.
- In 1999, Plug Power, a 50-50 joint venture between DTE Energy, the parent company of Detroit Edison, and Mechanical Technology of Latham, New York, began demonstrating its 7 kW Plug Power 7000 fuel cell. The power plant is expected to cost about $3,000 by the year 2003. The system will come with its own battery storage system.

- Between 1997 and 1999 the Sacramento Municipal Utility District (SMUD) has installed solar electric devices on the roofs of over 500 homes and businesses. Each system supplies about 75 percent of the home's annual energy needs. SMUD expects the solar electric roof shingles to be competitive with central power plants in three years.

The 1978 Public Utilities Regulatory Policy Act (PURPA) ushered in the era of independent power production. The average size of a new power plant dropped from almost 1,000 MW to less than 200 MW. Now an even more dramatic technological revolution may reduce the size of a new power plant by another 99 percent, from the megawatt to the kilowatt level, from the neighborhood scale to the small business and household scale.[1]

The technological revolution in power plant size has been accompanied by an equally dramatic but less heralded revolution in energy storage devices, such as flywheels and ultracapacitors, and in technologies, such as smart controllers, flexible dispatch algorithms and sensors capable of integrating millions of small power plants into a viable and stable grid system.

The age of the personal power plant is upon us. Today the United States is home to about 10,000 power plants. By 2003 there could be five times that many. By 2010 there could be over a million.[2]

The Electric Power Research Institute has estimated that the market potential of decentralized power is 5-50 percent of all new demand.[3] The Gas Research Institute has reached similar conclusions.

Thomas Edison would be delighted. The structure of the electricity system may be coming full circle, back to a time when the electricity industry was dominated by those who built and installed power plants in or near the customer. In the future businesses might sell and install power plants the way they now sell and install roof coverings, furnaces or central air conditioners.

Fig. 9. Size of Power Plant Needed for Different Enterprises

Application	Typical Size
Residential	3-7 kW
Small Business	6-50 kW
Large Retail	60-350 kW
Small Manufacturing	100-1000 kW
Large Manufacturing	1-100 MW

Source: Northern States Power Company

The era of decentralized power is here, and with it the possibility of designing an electricity system that treats people as producers as well as consumers, as sellers as well as buyers in the electricity marketplace.

Is Small Really Beautiful? The Economics of Decentralized Power

Decentralized power is wildly popular. But is it economical? That depends on how you do the math.[4]

Small power plants rarely generate electricity at a lower price than utilities can buy on the wholesale market. Today the price of wholesale power can be as low as 2-3 cents per kWh. Natural gas fueled micro-turbines generate power at 5-6 cents per kWh, wind turbines at 5-6 cents, small reciprocating engine-generator sets at 5-10 cents, fuel cells at about 6 cents and solar cells at 15-20 cents.

But using the current wholesale price as the basis for comparing the costs of on-site power plants is not appropriate. First of all, from the customer's vantage point, on-site power displaces retail-priced, not wholesale-priced electricity. Retail prices in many parts of the country are over 7 cents a kWh. In some cases they are above 12 cents.

Also, on-site power plants can fulfill several functions, thereby displacing other nonelectricity costs the homeowner would otherwise incur. A rooftop solar cell serves also as a shingle. The basement gas-fired power plant functions serves also as a furnace or boiler.

Finally, even when more costly, on-site power may still be desirable if it offers greater security or improved quality. These could become increasingly important considerations in the near future. For years utilities have put off buying new power plants. Reserve margins, the difference between total capacity and projected peak demand, fell from 25 percent in 1985 to 16 percent in 1998 and in the Midwest to only 13 percent. By 2003 it could drop to 10 percent or below in the west. "In five years, brownouts will be as common in the U.S. as they are in the Third World," predicts Randy MacCleary, vice president of the Distributed Power Group at Liebert Corp., a Columbus, Ohio, provider of a battery-based power-supply system.

"In five years, brownouts will be as common in the U.S. as they are in the Third World," predicts Randy MacCleary, vice president of the Distributed Power Group.

For Craig Schuttenberg, vice president of the Chicago consulting firm Energy Choices, "If power reliability at a business is a question you're much safer with your own power."[5] To Paul Colgan, director of public affairs for the Building Owners and Managers Association of Chicago, "It's something we feel that every commercial property owner should look at."[6]

It is not only the reliability of power but the quality of the electricity that concerns industries that rely on sensitive and sophisticated electronics. Power delivered over long transmission lines can degrade in quality (e.g. experience changes in voltage or frequency levels). Computers going down can result in hundreds of thousands, even millions of dollars of loss to an individual business every hour.

This concern led First National Bank of Omaha to install a 800-kW fuel-cell system in its 200,000 square-foot computer data center. The site provides 24-hour ATM, credit card and check-processing operations for many midwestern banks. The system costs $3,000 per kW, making it too expensive for typical commercial and industrial applications. The bank thinks the price is acceptable. "Roughly 45 percent of all computer outages are caused by power problems," says Art Mannion, executive vice president of Sure Power Corporation, the Danbury, Connecticut-based fuel cell supplier for the bank. "The project's key feature is its uninterruptible 'computer grade' power supply—voltage levels consistently within company manufacturers' specifications."

A deregulated electricity market may well make on-site generation, and storage, more valuable. Consider what happened in June 1998 in Illinois. The failure of one electricity marketer to deliver at a time when storms had knocked out several generators led the spot price of electricity to rise several hundred times. In this case the utilities swallowed the loss. In the future they could pass that cost onto their customers. If Illinois Power had done this, the average June electric bill for its customers reportedly would have reached $5,000.[7]

The current cost of decentralized power also doesn't reflect the economic benefits it offers to the electricity system as a whole.

In such volatile times, an on-site power plant may prove a wise investment. Indeed, the June price hike apparently spurred a dramatic increase in power plant orders. According to the *Wall Street Journal*, the price of power equipment, which had plummeted nearly 50 percent since 1993, rebounded 10-15 percent after the events of June.[8]

The current cost of decentralized power also doesn't reflect the economic benefits it offers to the electricity system as a whole. Today getting the electricity to us and managing the complexity of the electricity system in many cases costs more than it does to generate the power in the first place. According to the Pacific Gas and Electric Company and the Energy Information Administration, some utilities spend $1.50 to distribute power for every $1.00 they spend producing it.[9] Decentralized power can cut these costs.

Another benefit that distributed generation provides to the electricity system as a whole is a reduction in transmission line losses. According to energy consultant Tom Starrs, 3 to 10 percent of electricity carried across transmission lines is lost, depending on the distance and other factors.[10] This obviously has both environmental and economic consequences.

As homes install more computers, bigger televisions, ice-making refrigerators and central air conditioners, their demand eventually will exceed the peak carrying capacity of the existing electricity distribution lines. Instead of expanding the capacity of the lines, in the future utilities may find it cheaper to install power plants at the customer's site.

This is exactly what happened in New York City, where Consolidated Edison's distribution system could not meet the growing electric needs of the 150-year-old Central Park police precinct station. The cost of upgrading the lines was estimated at $ 1.2 million, and construction was expected to exact a long and heavy toll on the Central Park landscape. Instead, a single PC25 fuel cell—about the same size as a large garden shed—now supplies 200 kW of electricity to the station. It also provides power to recharge nonpolluting electric vehicles used by police to patrol the park. The station is now completely separate from the Con Ed grid.[11]

Southern California Edison found that a 150-kW solar-cell system can deter replacement of a 4-kW distribution line into a residential neighborhood, avoiding the disruption that comes from tearing up residential streets, the loss of customer good will, and nearly $1 million in costs.

Fig. 10. The Economic Benefits of Decentralized Power

Electric system benefits	
Substation deferral	0.16-0.6 cents per kWh
Transmission system losses	0.2-0.3 cents per kWh.
Transmission wheeling	0.28-0.71 cents per kWh.
Distribution benefits	0.067-0.17 cents per kWh.
Enhanced reliability	1.0 cents per kWh.
Total	1.7-2.8 cents per kWh.
Societal environmental benefits	
SO_2 emission offset	0.15-0.45 cents per kWh.
NOX emissions avoided	0.01-0.3 cents per kWh.
CO_2 emissions	0-1.5 cents per kWh.
Total	0.16-2.25 cents per kWh.

Source: DPCA, 2000

The benefits of distributed generation may be best realized by rural cooperatives. The nation's 900 coops own nearly half of the country's distribution lines—enough wire to circle the earth 80 times. Nearly half of these lines were installed over 40 years ago and soon will require upgrading. Rural coops have very low customer densities, and thus would have to spread out the cost of distribution upgrades over a relatively small number of customers. A recent study found that instead of upgrading many of these wires, it would be cheaper to install solar cell and propane co-generator hybrid systems at dispersed sites. Solar cells are well suited for summertime use when air conditioning demand is high, and in the winter the waste heat from propane generators can be captured to warm homes and businesses.

Distributed generation also can substantially reduce the financial risk associated with central power plant construction. A large central power plant can take 3-6 years to come on line. That requires accurate forecasting. Decentralized power plants come on line in a few months, adding small increments of capacity to the system. Ultimately, every increase in demand (e.g. another business opening or an office building going up) could bring with it its own increase in supply (e.g. a basement power plant or a rooftop solar cell system), eliminating any risk of over- or underestimating demand.

The Distributed Power Coalition of America (DPCA) has quantified the many benefits of decentralized power plants. Its conclusion? Decentralized capacity reduces system costs by 1.7-2.8 cents per kWh and offers environmental benefits worth 0.16-2.25 cents per kWh. Total benefits could be over 5 cents per kWh, making virtually all decentralized power plants competitive with central power plants.

Indeed, Gerry Runte, director of the Eastern Regional Office of M-C Power Corporation, a builder of fuel cells, insists that in the future "any delivered cost of electricity between 5 and 8 cents/ kWh will indeed be quite competitive, even in the cheaper world of a deregulated future, post stranded costs and post wholesale price reductions."[12]

The attractiveness of decentralized power has led at least 100 utilities to initiate formal studies of their costs and benefits. Several industry and utility groups have been established to actively promote decentralized power.[13]

If dramatic breakthroughs occur in the cost of storing electricity, buildings, businesses and farms could completely uncouple from the grid system. But that is unlikely. Autarchy demands that the on-site power plant be sized for the building's peak load rather than its average load. Peak demand can be two or three times average demand. Autarchy also requires redundant capacity, so that when one power plant is down for maintenance or other reasons, another will be available.[14]

More likely is for us to become, in Alvin Toffler's term, "prosumers." Buildings will become self-reliant, not self-sufficient. Building owners and independent suppliers will sell and buy power via an electric grid system that functions like a giant marketplace, with prices varying by time of day and type of contract.

Decentralization of power production, however, will not occur inevitably nor will it happen overnight. Indeed, the sudden appearance of looming shortages of electricity in late 2000 could provide a dramatic spur to small power plants, or retard their appearance. Utilities stopped building central power plants in the late 1980s, when recession slowed electricity demand and independent power producers were gaining maturity. In the 1990s most utilities didn't build large central power plants because they were waiting for deregulation. Indeed, even in those states that haven't deregulated, utilities no longer see themselves as in the power plant building business.

Buildings will become self-reliant, not self-sufficient. Building owners and independent suppliers will sell and buy power via an electric grid system that functions like a giant marketplace.

Thus in 2001 there is a need for electricity, and a huge supply backlog of large central power plants that need new transmission lines to reach their distant customers. If the transmission capacity is there, and these plants get built, it is possible that by 2005 or so, we may be in a situation where we are in surplus once again, and decentralized power plants again must struggle uphill for a market.

To literally bring power to the people, a nationwide effort is needed to fashion rules that will accelerate the development and deployment of distributed generation technologies.

Designing Rules to Decentralize Capacity

Only recently has decentralized power become a part of the state-level debates over electricity deregulation. Texas, which passed its restructuring legislation in 1999, has most aggressively promoted distributed generation as a solution to capacity strains in a deregulated environment. The Texas Public Utilities Commission has been given the authority to write the rules that will govern distributed generation (DG) in the Longhorn State.[15]

The PUC's most noteworthy decision thus far has been its approach to the costs of pre-interconnection studies. Because distributed generation benefits the whole electric system, the commission decided to spread the costs of interconnection studies for small customer/generators (less than 500 kW and that export no more than 15 percent of the total load of a single neighborhood distribution line) among all ratepayers. The commission decided that self-generators larger than this threshold should bear the costs of any interconnection studies.[16]

Texas is currently developing a system for certifying decentralized power plants, preparing for the day when such units will be standardized off-the-shelf items.[17] Such certification procedures are essential to encourage equipment manufacturers to mass-produce DG equipment, which is necessary to bring down costs. Currently, most utilities have their own procedures and standards that often require site-specific engineering, which increases the cost of each DG unit installed.

California got an earlier start with its investigation into distributed generation, but its progress has been slow. In early 1999 the CPUC, in association with the California Energy Commission and the California Electricity Oversight Board, launched a formal investigation of how distributed generation will affect the competitive viability of utility distribution companies. A new rulemaking was opened by the CPUC in 2000 to investigate and remove inappropriate barriers that utilities have erected to prevent the deployment of DG resources. Another priority is the development of interconnection standards.[18] In February 2001, the Omnibus Distributed Energy Resources and Clean Electricity Act of 2001 was introduced in California. The bill provides a comprehensive framework for the use and management of distributed energy resources and accelerates the implementation of that framework.

Illinois and New York have solicited comments and held workshops in an effort to determine how they want to handle distributed resources in the context of a deregulated industry. They and other states will likely proceed in a way similar to Texas and California.

Up to this point state regulatory agencies have played the primary role in fashioning the new small-scale power rules, and they will continue to do so. However, local and state governments can also encourage the growth of decentralized power by updating land-use ordinances and building codes and standards to remove any obstacles or barriers to on-site power.

Rule #1: Include decentralized power as an element in local and regional plans

If a proliferation of small-scale power plants serves the interests of the general community, cities and counties should include this as an element in their general plans and zoning ordinances. San Diego's regional energy plan includes Implementation Measure 14: Small Scale Distributed Power Generation. "The objective of this measure is to increase awareness of distributed power generation technologies generally; to ensure that institutional and legal barriers do not impede their development, e.g. siting standards, and to encourage their use when meeting small increments of the region's electric needs…"[19] This is a perfect example of how a local government can begin to encourage local power production.

Local governments must also redesign their building and electrical codes to remove obstacles to on-site power. Currently, these codes provide guidelines for the installation of appliances. They will need to be updated to include guidelines for the installation of power plants.[20]

Rule #2: Encourage utilities to sell their power plants

For the foreseeable future—perhaps forever—distribution and transmission will remain monopolies.[21] Therefore utilities that own the power lines will and should continue to be regulated. But there is an inherent conflict of interest in utilities owning the distribution system and also owning power plants that will compete with other independently owned power plants. Indeed this has already occurred where divestiture is not mandated.[22]

Both the Federal Trade Commission and the Department of Justice agree that the only long-term solution to the abuse of market power by existing utilities is to force them to divest their power plants. "As a general proposition," argues the FTC, "we have found that structural remedies, such as divestiture in merger cases, are the most effective and require the least amount of subsequent monitoring by government agencies. The effectiveness of structural remedies lies in the fact that they directly alter incentives. Behavioral remedies, in contrast, leave incentives for discriminatory behavior in place and impose a substantial burden on government agencies to monitor subsequent conduct."[23]

Many states undertaking retail competition, including Maine, Massachusetts and Rhode Island, do require at least a partial divestiture of power plants by existing utilities. California requires that utilities sell off at least 50 percent of their generation capacity.[24] In Connecticut utilities must sell their nuclear plants before they can receive full recovery of their uneconomic costs. On the other end of the spectrum, in Illinois divestiture is not allowed.

The burden of proof should be on those who oppose divestiture. If utilities are not going to structurally separate their generation and sales divisions then regulatory agencies must develop a clear and effective incentive program that will achieve the same purpose of enabling decentralized power.

Rule #3: Tie utility revenues to lowered costs, not increased sales

In a competitive environment, divestiture does not do enough to remove utility incentives to discourage distributed generation. Under volumetric ratemaking, distribution companies (or "discos") make money by transferring as much electric power as possible over their wires to final customers. Every on-site power plant installed, therefore, equates to lost utility revenue.

"From the utilities' standpoint, the first thing they think of is that (distributed generation) is a threat to their load," says Ritchie Priddy, manager of the distributed energy resources program at Louisiana State University. "They have this attitude that I call 'love it to death' syndrome; saying they support it, but when it's in their own back yard they are less than aggressive about it."[25]

Utilities are also wary of self-generation because of its potential to perpetuate itself. As more and more customers become self-generators, utilities are forced to raise rates because they have to spread out their capital costs over fewer customers. Higher rates encourage more self-generation, which results in still lower utility revenues.[26] Utilities are therefore likely do everything in their power to slow the development of decentralized capacity—unless a new regulatory scheme can be adopted that prevents utilities from losing significant revenue from self-generation.

Oregon may have found the solution. In late 1999 that state's PUC adopted a performance-based ratemaking tariff for PacifiCorp. The plan applies a "revenue cap" to each customer class. If actual sales revenues exceed the predetermined cap in any of the classes, the extra is set aside in a balancing account. The following year, the balancing account funds are given back to the utility if sales were lower than projected, and given back to customers if utility sales were higher than anticipated. In this way, the utility's revenues are disconnected from the amount of electricity it distributes, thus eliminating any reason to discourage customer generation. Ralph Cavanagh of the Natural Resources Defense Council calls Oregon's plan "a wonderful new regulatory precedent."[27]

Rule #4: Get the prices right: value decentralized power
With or without divestiture or performance-based ratemaking, regulatory authorities should require utilities to establish a transparent pricing system so that consumers can easily understand the elements that make up their electric bill.

Utilities should unbundle the customer's bill. This step can be taken in states that are not embracing retail competition as well as those that are. This means that the customer "sees" the many elements of an electric bill: energy charge, demand charge (cost of providing electricity during peak hours), local wires improvement cost, stand-by charges, line losses, reliability values, voltage regulation, power factor maintenance, etc.[28]

Transparency of pricing may not be enough. As Ralph Cavanaugh writes, "the regulatory objective is to break the link between the profitability of the distribution monopoly and the amount of kilowatt hours flowing over the distribution wires..."[29] The objective of the future electricity system should not be to expand traffic but to optimize efficiency. Sometimes that will mean expanding traffic, but often it will mean expanding on-site generation or dispersed storage, or improving efficiency.

Regulatory commissions must quantify the benefits of dispersed power (and efficiency and storage) and develop pricing mechanisms and incentive programs that take these into account. In doing this, they can learn from the experiences of regulatory commissions who quantified the "avoided costs" related to energy efficiency investments.

Pricing will influence the shape of the future electric system and perhaps the size of the future power plant. As Leonard Hyman and Marija Ilic note, "When grid operators set a price and decide which transactions will take place, they influence the value and location of power. We're not talking about arcane, opaque, academic engineering exercises. We're talking about real money."[30]

At the transmission level, FERC has been neutral on the pricing strategies, so long as the system adopted is applied equally to all suppliers. The existing transmission system uses "pancake" pricing. The marketer is charged an additional transmission fee each time the electricity crosses a utility's jurisdiction. Since at present transmission systems are under local-utility ownership, this means the cost of transporting electricity long distances is quite high. Such a pricing system favors local generation.

When grid operators set a price and decide which transactions will take place, they influence the value and location of power.

The new regional transmission authorities emerging under FERC rules are adopting three types of pricing.

One is "postage stamp" pricing, so named because it is similar to the way the U.S. mail system is priced. The transmission charge is the same whether the electricity goes a few feet or a few hundred miles. This system is the most favorable to long distance electricity traffic.

The second is based on miles traveled. This favors more dispersed, local generation.

The third major system is called location based marginal pricing (LBMP), or congestion pricing. Here the price of transmission is based, not only on the cost of the transmission lines, but on the availability of capacity. As was pointed out before, electricity rarely travels along the contract path. The owner of a power plant in Montana might sell power to the owner of an office building in Seattle, but the electricity generated in Montana may well go to Los Angeles because of the physics of electric flows.[31] One analysis concluded, "a power transfer from Indiana to New Jersey would produce flows over the lines of more than 20 different utilities and power pools. Less than half of the transferred power would flow over what would appear to be a relatively direct route."[32]

LBMP, unlike any other pricing system, takes into account these loop flows of electricity caused by the varying congestion levels of transmission lines. Pennsylvania, New Jersey and

Maryland's transmission pool adopted LBMP in 1998. It computes the marginal cost of transporting electricity at 1,600 locations in its region. When congestion occurs, prices on the customer side of the transformer can be considerably higher than prices on the other side. This system favors more dispersed generation.

The Texas PUC has endorsed the principle of congestion-based pricing in its investigation into distributed generation by ruling that distribution charges should be near zero for areas with excess distribution capacity but should be high in areas with congested distribution facilities. "Making customers pay the full incremental cost of distribution," *Public Citizen* argued in its comments to the Commission, "will provide an incentive to make more rational decisions about the deployment of distributed resources."[33]

In October 2000, LBMP received an unofficial endorsement from FERC head William Massey at an Edison Electric Institute meeting. "I don't think it's any secret that I find great value in the locational marginal pricing, or LMP, model," he said. "LMP sends the correct price signals needed for optimal use of existing generation and transmission resources and also encourages efficient siting of future generation and transmission expansion. And I do not think I'm alone. I think the commission tilts toward LMP, perhaps even looking on it as a presumptive favorite."[34]

Pricing strategies that reflect the true costs of distribution will encourage the siting of distributed generation resources when and where they are most needed.

Rule #5: Require the consideration of distributed generation as an alternative to distribution line extensions and replacements

Arizona and Colorado, as a matter of regulatory policy, require utilities to compare the cost of extending a distribution line with the cost of a PV/hybrid system (photovoltaic system with a gas generator backup) to serve new customer load.[35] Other states should adopt a similar policy that requires consideration of distributed generation as an alternative to upgrading or replacing distribution systems. If a state considers distributed generation to be an especially high priority (especially environmentally benign distributed technologies), it could give those technologies a five or ten percent price advantage. For example, if the cost of installing fuel cell powered cogenerators was only five percent more expensive than upgrading the distribution lines, the utility could be required to install the fuel cells.

One way to give these environmentally benign power sources credit is to require power suppliers to include in their price the cost of pollution from their power plants.

In California, where utilities are not required to consider distributed generation as an alternative to their traditional wires solutions, they don't. Utility planning engineers, reports the California Energy Commission, typically have little familiarity with on-site generation, and will usually defer to what they know works, even if it doesn't work best.[36]

Rule #6: Require electricity suppliers to include pollution costs in their prices

In the late 1980s and early 1990s about half a dozen state regulatory agencies quantified the environmental costs of power generation and began to include these costs when evaluating competitive bids for new power projects. In a competitive marketplace, many are concerned that pollution will no longer be taken into account since clean energy suppliers will have to charge more. Increasing numbers of suppliers are marketing "green" power. This will be discussed in chapter 5.

Zero emission power plants tend to be decentralized power plants, including on-site fuel cells, solar electric devices and wind generators. One way to give these environmentally benign power sources credit is to require power suppliers to include in their price the cost of pollution from their power plants.

Today states generate about $17.5 billion in tax and fee revenue from electricity generators. In a competitive market, some power companies that pay these taxes could see themselves at a competitive disadvantage with providers that are not subject to the same tax burdens. Thus states that are embracing retail competition have also changed their electricity tax structure. Many have adopted a kWh charge for all electricity sold. They could easily convert this into a pollution tax per kWh sold.[37]

Net metering allows on-site generators to, in effect, receive the retail price for the electricity they would otherwise have purchased.

Adopting such a tax would significantly raise the cost of fossil fueled electricity, encouraging the use of more decentralized renewable energy sources like direct sunlight and wind. For example, $17.5 billion translates into an $85 tax per ton of carbon emitted. This would raise the price of coal by more than 1.5 cents per kilowatt hour, making wind energy competitive without the need for a federal tax incentive.

Rule #7: Require net metering

The rules proposed above will encourage the rise of more dispersed power plants, but as the Office of Ratepayer Advocates of the California Public Utility Commission has stated, "It is vitally important to distinguish between distributed resource applications on the supply side vs. the customer side of the meter."[38] They urge policymakers to adopt rules that would encourage on-site power plants. One such strategy is called "net metering." More than two dozen states have already enacted net metering statutes.

Net metering laws allow a customer's meter to essentially run backwards. In most states, at the end of the month the customer pays (or receives) the net difference between the amount generated and the amount consumed. Some states allow customers to credit any excess in a given month against the following month's bill.

Net metering allows on-site generators to, in effect, receive the retail price for the electricity they would otherwise have purchased. Any excess electricity is usually sold to the utility at the same rate the utility would have to pay to buy that electricity on the open market (known as the utility's avoided cost). Some states, like Indiana, do not require the utility to pay anything for the excess power. Two states, Wisconsin and Minnesota, require the utility to pay the retail price for excess power (subject to certain conditions and system size limitations).[39]

Some states offer net metering only to residential customers. Some, like California and New York, limit it to solar energy devices.[40] Others, like New Mexico and Connecticut, offer net metering to all customers and gas-fueled devices as well as renewable energy technologies are eligible.[41]

All states impose a size limit for net metering eligibility. California and New York impose a limit of 10 kW, Arizona 100 kW and New Mexico the highest in the nation at 1000 kW.

Net metering laws provide a significant incentive to on-site electric generation. They are also controversial because they, in effect, allow an on-site generator to pay nothing for the privilege of interconnecting to and using the electricity distribution system. A customer could, in theory, end up at the end of the year having not paid the utility a cent. Clearly if everyone installed a home power plant that kind of tariff would result in the bankruptcy of the distribution system.

Thus net metering statutes should be viewed as temporary strategies to quickly build independent power capacity. They reward the pioneers. Most states impose some limit to the number of net metering customers. Nevada, for example, imposes a ceiling of 100 customers for each of its utilities. California puts a limit of 0.1 percent of peak demand (equal to 53.3 MW). Vermont's limit is 1 percent of peak demand.

Net metering is currently under attack. In August 1999 an Iowa District Court ruled that federal law preempts Iowa's ability to impose a net metering law on utilities. The court ruled that net metering violated PURPA by requiring a utility to purchase electric power at a rate greater than its avoided cost rate. The court also found that net metering requires utilities to purchase power that becomes commingled with other energy that is sold in interstate commerce, thus violating FERC jurisdiction over the setting of wholesale rates, which is required by the Federal Power Act.

The Iowa Utilities Board (IUB) and the Office of Consumer Advocate have appealed the decision. The IUB, along with other state commissions, argues that net metering "relates to a utility's metering and billing practices, which fall squarely within state regulatory jurisdiction over retail practices of electric utilities and are not preempted by FERC rulings or other federal law."[42]

RULES FOR DECENTRALIZING CAPACITY

1. Include decentralized power as an element in local and regional plans
2. Encourage utilities to sell their power plants
3. Tie utility revenues to lowered costs, not increased sales
4. Get the prices right: value decentralized power
5. Require the consideration of distributed generation as an alternative to distribution line extensions and peplacements
6. Require electricity suppliers to include pollution costs in their prices
7. Require net metering

CHAPTER FOUR

Assuming Authority: Taking Control of Our Electricity Systems

One-third of us own our electric companies, either directly as members of 900 cooperatives, or indirectly as citizens of the 2,100 municipalities that own their utilities. Two-thirds of us are customers of 240 investor-owned utilities (IOUs).

A third type of ownership structure is the federal power authority. These are publicly owned but their chief executives and directors are recommended by the President of the United States and confirmed by the U.S. Senate. Decisions are therefore made at such a remove that it's hard to view them as customer-owned. They generate a small portion of the nation's electricity but can dominate some regions. For example, more than half the electricity sold in Washington state comes from the Bonneville Power Authority, which also owns 75 percent of all transmission lines in the Pacific Northwest. The Tennessee Valley Authority dominates in its region.

Fig. 11. Composition of the Electric Power Industry in the United States

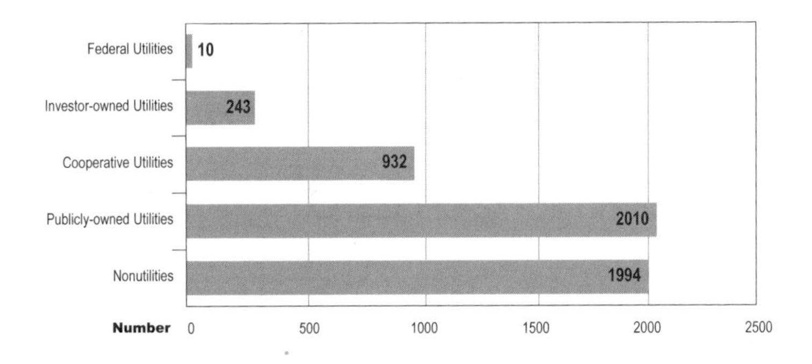

Notes: Power marketers, Puerto Rico, and U.S. Territories are not included. Nonutilities represent the number of generating facilities, as these facilities are generally incorporated, and each is required to file Form EIA-867.

Source: Energy Information Administration, Electric Power Annual 1996 Volume II, February 1998

Ownership patterns vary dramatically by state. Nebraska's electricity system is 100 percent publicly owned. Hawaii's is 100 percent investor-owned. Only about one in thirty people in New York, New Jersey and Pennsylvania are served by customer-owned electric utilities (COUs). The ratio rises to about one in ten for people living in New England. In midwestern states from North Dakota to Missouri the proportion rises to four in ten, and in the southern states of Alabama, Tennessee and Mississippi two-thirds of the electric utilities are customer-owned.

Which organizational structure for utilities is better—customer-owned or investor-owned? The answer has been vigorously contested for over 100 years. Both sides have data to justify their position. Municipally owned utilities, or "munis," point to their overall 18 percent price advantage. IOUs respond that this is a result of munis' ability to use low-cost tax exempt bonds and their access to cheap federal power. Munis reply that IOUs benefit even more handsomely from federal tax laws and claim that even without the use of tax exempt bonds, municipal utility rates are still 10 percent lower.[1]

Cooperatives argue that where their rates are higher it is not because of management inefficiencies but the fact that they serve fewer customers per mile and their customers are primarily high-cost residential customers.[2]

The most important distinction is that customer-owned utilities are inherently more democratically governed, closer to their customers and more responsive to them.

The debate is useful, but 100 years of actual experience teaches us that both ownership structures are efficient, reliable and innovative. The most important distinction is that customer-owned utilities are inherently more democratically governed, closer to their customers and more responsive to them. The federal government has recognized these differences, and since the 1930s has established policies to nurture customer-owned electric utilities. Will we continue to do so in the age of deregulation?

Five Key Differences Between IOUs and COUs

Customer-owned and investor-owned electric utilities differ in at least five fundamental ways.

1. Customer-owned utilities often are self-regulated. All states regulate investor-owned utilities in return for granting them monopoly status. Since customer-owned utilities are governed by their customers, most states allow them to regulate themselves. Only 41 states have meaningful jurisdiction over municipal utilities; 31 regulate rural electric cooperatives.

This exemption continues into the age of deregulation. States restructuring their electric systems often exempt customer-owned utilities, giving each one the option of embracing retail competition. Congress has exempted federal power agencies, municipal utilities and most cooperatives from the provisions of the Energy Policy Act of 1992.

2. COUs are subordinate to state authority in their pursuit of nonelectric activities. IOUs are not.

In an odd twist, while customer-owned utilities tend to be exempt from a great deal of direct federal and state regulation, they are more subordinate to federal and state regulations when it comes to their pursuing nonelectric activities. For example, in the information age, any enterprise that connects all homes and offices in a billing and metering relationship can easily add more services. More and more utilities are doing so. More than 150 investor-owned electric utilities now sell long distance or local phone service, cable television, home security systems or internet connections.[3]

Customer-owned electric utilities are moving into these new business sectors as well. In Iowa alone, more than 30 cities have voted to have their customer-owned electric companies provide telecommunications services.[4] In Kentucky, 11 municipally owned utilities have banded together and formed the Northern Kentucky Telecommunication Authority (NKTA). In 1986, several hundred electric cooperatives formed the National Rural Telecommunications Cooperative (NRTC). In 1993, NRTC participated in the launch of a new satellite television system, the direct broadcast system (DBS). According to UT Digest Magazine, 65 cities, including Tacoma, Washington, and Boulder, Colorado, have built high-tech telecom networks and are offering services through their municipal utilities.[5]

Private corporations, like investor-owned utilities, are virtually immune from state regulation over the kinds of services they can offer. But public corporations, like municipally owned utilities, are, according to the U.S. Supreme Court, "creatures of the state." Thus state legislatures can prohibit or restrict the entry of customer-owned utilities into services like telecommunications. Several have, including Texas, Missouri and Virginia.

Rulings issued by the Federal Communications Commission (FCC) reflect this dual approach to IOUs and COUs. In 1996 Congress enacted the Telecommunications Policy Act, which opened up that industry to competition. Section 253 clearly states, "no state or local statute or regulation, or other state or local legal requirement, may prohibit or have the effect of prohibiting the ability of any entity to provide any interstate or intrastate telecommunications service."

But in 1995 the Texas Public Utility Regulatory Act prohibited municipalities and municipal electric utilities from furnishing "telecommunications services" either directly or indirectly. The FCC declined to preempt the Texas legislature. Missouri and Virginia followed Texas' lead. But when tiny Hill City, Kansas decided to forbid a private company from offering services in its territory and instead to build its own customer-owned telecommunications system, FERC ruled that the city was pre-empted from doing that by the federal law. It turns out that when Congress said "any entity," it really meant any "private, for-profit entity."

3. IOUs own most of their generating capacity; COUs do not.

Only a handful of IOUs don't own any generating capacity, while two-thirds of municipally owned utilities own virtually no power plants. Cooperatives are in-between. Distribution cooperatives own no generating capacity but their collectively owned generation and transmission cooperatives (G&Ts) provide about half of the electricity purchased by distribution cooperatives.

In many states, this distinction is eroding as both customer-owned and investor-owned utilities increasingly become "wires" or distribution utilities. IOUs are selling much of their generation capacity, either voluntarily or because of state divestiture requirements. A growing number of distribution cooperatives are also uncoupling from their G&T coops and buying their electricity directly on the wholesale market.

4. COUs are place-based. IOUs are increasingly national and international conglomerates.

The ownership structure of customer-owned utilities makes them inherently local. It is hard to conceive of the Los Angeles Department of Water and Power buying Seattle's municipally owned utility. But IOUs, once headquartered in the area where they sold their electricity, are becoming increasingly disconnected from their customers.

The ownership structure of customer-owned utilities makes them inherently local. It is hard to conceive of the Los Angeles Department of Water and Power buying Seattle's municipally owned utility.

Since 1992, about 100 utility mergers and acquisitions have taken place. About half of the entire asset base of private utilities has been the subject of merger activity in the last four years.

The rate of utility mergers isn't the only thing increasing—the scale of these energy marriages is also skyrocketing. Between 1992 and 1998, only four mergers were completed in which the combined assets of the companies exceeded $10 billion. But of eight mergers completed in 1999 or 2000, each has combined assets greater than $10 billion.[6]

Utility executives justify these mergers as a way to lower costs, but the estimated savings rarely exceed 2 percent, and most of this is achieved by reducing jobs. No analysis has ever been done on the impact of job reduction on, for example, response time to outages or on-going maintenance.

Behind the frenzied merger activity stands a player whose objective only vaguely relates to lowering prices or increasing share prices: the investment banker. "Merchant banks are estimated to have initiated upwards of 66% of all M&A activity," one trade journal declared. Banks advising PacificCorp in its bid for the Energy Group accrued fees in excess of $100 million and the deal is not done. Thomas Hamlin, a utilities analyst with Wheat First Union in Richmond, Virginia, observed, "Everyone is up for sale. Everyone is going to be bought."

Mergers are also occurring among some customer-owned utilities. In 1998, cooperative and municipally owned electric utilities were allowed to merge without federal approval. At least nine mergers among co-ops have occurred in Iowa and the Dakotas alone since then.

When these mergers occur, they are almost always neighborly affairs, resulting in what Mark Glaess, manager of the Minnesota Rural Electric Cooperative Association, calls a "virtual merger."[7]

For example, Lake Country Electric Cooperative in Minnesota was created after members of its three original coops voted overwhelmingly to merge. The new co-op has one general manager instead of three. But none of the three former co-op offices were closed. The general manager works in Grand Rapids. the manager of engineering and operations works in Virginia, and the head of marketing is in Kettle River.

Mergers among cooperatives tend to maintain local representation. Thus when the Minnesota cooperatives of East Central and North Pine Electric merged, the board of directors was cut from 21 members to 14 and redistricted by population.

5. IOUs and COUs have dramatically different decisionmaking processes and corporate goals.

The most fundamental difference between IOUs and COUs, of course, is in their decisionmaking structure and legal missions. IOUs are profit-making; COUs are not-for-profit. Investor-owned utilities, by law, are required to maximize the value of their shares. Customer-owned utilities are required to maximize the value of their service.

The governance structure among IOUs is quite similar: those with the most stock have the most votes. Cooperatives are governed, directly or indirectly, by the principle of one customer, one vote. Municipal electric companies are usually governed directly by the city council. Sometimes there is an outside board of directors and sometimes that board is directly elected by the citizens of the area.[8]

Recently, the structural and mission differences between IOUs and COUs were highlighted by the American Public Power Association (APPA). IOUs have argued, with some success, that COUs shouldn't be allowed to issue low-interest tax-exempt bonds to finance their facilities if they are competing in a deregulated market, because this low-cost money gives them an unfair advantage.

APPA responded that private power companies should not be allowed to compete in a deregulated market if they do not adopt the same democratic decisionmaking processes as COUs. IOUs, APPA insisted, should have to hold open meetings, make their records public and hold public hearings on their budgets. They should have to allow the public to elect their board and recall their CEO if necessary. The IOU should have to hold a referendum to approve the construction of new power plants, and gain public approval for the use of eminent domain authority.[9]

Illustrating the Differences Between COUs and IOUs

Municipally owned utilities can and have been sold to investors. The process requires a public vote. Recently, two mayors recommended that their citizens act like wise investors and reap a windfall profit by selling their municipally owned systems.

As Mayor Scott Maddox of Tallahassee tells the story, at a strategic planning session he asked, "Why are we in the electricity business? Is it to provide electricity? To preserve energy? Or is it for the money? Everyone agreed it was for generating revenue." Today power plants and distribution systems are selling at a premium price. It is a seller's market. A reporter sums up the mayor's reasoning, "If a private suitor was willing to pony up three times the book value, they figured it would outweigh the local control concern."

The citizens of Tallahassee adamantly disagreed. They thought that local control outweighed a short term profit. The city council voted down the mayor's proposal.[10]

In Memphis, Mayor Willie Herenton tried to sell the Memphis Light Gas and Water Division. In a sellers market, he reasoned, the sale could bring in $600-800 million, which might generate $70 million a year for the city, three times what the utility was currently producing in revenue for the city. Mayor Herenton insisted, "I don't know any corporation that would have an asset like the city of Memphis has on its books and not try to secure a bigger dividend from that asset."

Some 74 percent of those polled by the local newspaper, *Commercial Appeal*, preferred local control. The mayor backed down.

The citizens of Tallahassee thought that local control outweighed a short term profit. The city council voted down the mayor's proposal.

To see how a state approaches deregulation when the customers own the entire electricity system, we might examine Nebraska, the nation's only 100 percent public power state. Its retail customers are served by more than 200 publicly owned distribution companies. They purchase their power wholesale from a handful of public power districts that also own the transmission grid.

Nebraska's rates are among the lowest in the nation. The state's public power districts have maintained an adequate supply of generation and transmission in the face of growing demand. Because Nebraska's electric system is owned by its citizens, there are no competing demands from ratepayers and shareholders; the ratepayers are in fact the shareholders.

After an exhaustive three-year analysis, the Nebraska legislature passed a unique restructuring law. Instead of setting a date by which the state had to institute retail electric competition (an approach all other states with restructuring laws have taken), Nebraska lawmakers, by a 47 to 0 vote, established a set of conditions that had to be met before the state could embrace competition.

The most important of these conditions is the existence of a viable regional wholesale market, complete with a functioning regional transmission organization and sufficient transmission capacity. In California, of course, none of these conditions exist five years after its restructuring bill was passed. Secondly, before deregulation could be implemented, and to insure cost benefits, Nebraska's wholesale electricity prices must rise above those of neighboring states.

Ironically, if Nebraska was to "deregulate" its electric power industry, it would require more regulation, not less. Currently, because all utilities are locally owned and controlled, there is no need for a state regulatory body. If retail competition were introduced, a statewide agency would need to be created to monitor the market and resolve disputes over billing and service.

Marketing Community

In the coming battle for the hearts and minds (and dollars) of electricity customers, customer-owned utilities are weighing in with marketing efforts that promote a sense of place and ownership. Touchstone Energy is a national alliance of over 400 cooperatives. Officially formed in April 1998, Touchstone's logo features three figures holding hands with the tagline "The Power of Human Connections." The Touchstone alliance, which serves more than 11 million customers nationwide, has also set up its own marketing agency to sell electricity to new customers.

The California Power Network (CPN) was formed in the fall of 1997 by 21 munis in California who pooled their resources to advertise the benefits of municipal power. In March 1998, CPN launched a campaign to tout the benefits of community-owned electric power, a campaign timed to coincide with the opening of competition in California. Ads stressed munis' low rates and local roots.

Fashioning the Rules for Assuming Authority

The most direct way to ensure that authority is vested with the citizenry is through customer ownership, as we have seen. But in many cases public ownership may not be possible or practical. In these instances there are a number of other strategies that communities and policymakers can embrace that would make our electric power system more democractic, responsive and, in the end, possibly more equitable and environmentally benign.

Rule #1: Declare a moratorium on large energy mergers at the state and federal level.

As was noted above, a frenzy of mergers and acquisitions has been sweeping through the electricity and natural gas industry. There is a sense by utilities that in the era of competition they must "bulk up." Says Fred D. Hafer, chairman and CEO of GPU, "There seems to be a consensus building that probably 10 million (customers) is big enough (to compete)..."[11] That is larger than any existing electric utility.[12]

Mergers by definition move those who make the decisions further away from those who feel the impact of such decisions. Currently there are over 200 investor-owned utilities in the U.S, many still headquartered near their customers. None has more than a 3 percent national market share. But many observers, including FERC Commissioner William Massey, expect that within a decade only a handful or so will remain. "I think you'll eventually have a half dozen or so big generating companies and a dozen or so big transmission companies," he says.[13]

Such prognostications are already proving correct. By the end of 2000, the number of IOUs that own generation capacity was an estimated 141—down from 172 in 1992. Consolidation is even more apparent when capacity is aggregated by holding companies. In 1992, there were 70 electric holding companies owning 78 percent of the IOU-held generation capacity. By the end of 2000, the number of electric holding companies has dwindled to 53, the ten largest of which own fully 50 percent of the total IOU-owned capacity.[14]

These new utility behemoths hearken back to the pre-regulation days of the 1920s and 1930s, when ten holding companies controlled three-fourths of the nation's electricity business, with J.P. Morgan alone owning nearly half.

To date FERC has taken a very lax attitude toward mergers. Indeed, from January 1998 through January 1999 it disapproved only one proposed merger. When Long Island Lighting Company proposed to merge with Brooklyn Union Gas, evidence showed that together they would control 47 percent of the aggregate natural gas pipeline capacity into Long Island. FERC saw no problem. In the merger case involving Enova and Pacific Enterprises, evidence showed that Southern California Gas Co delivered natural gas to 60 percent of the generating capacity that generally was available to supply electricity to the Southern California market. FERC approved the merger.[15]

Even while the FERC works out the rules for wholesale wheeling and the governance structures of regional transmission systems, it is approving massive mergers.

Even while FERC works out the rules for wholesale wheeling and the governance structures of regional transmission systems, it is approving massive mergers. This has led Joseph Klein, head of the U.S. Department of Justice Antitrust Division, to call for a moratorium on further merger approvals. "A moratorium" he said, would "postpone making difficult competitive evaluations for a brief period until we have developed a market-based history." The American Public Power Association and the National Rural Electric Cooperative Association petitioned FERC in 1998 to declare a moratorium on future mergers of large utilities.

Merger proponents argue that they improve efficiency, but the utilities' own data indicates that at best, the improvements are meager. In March 1999 Northern States Power proposed to merge with New Century Energies, a merger that would create a new utility (called Xcel Energy) serving 3.5 million electric and 1.5 million natural gas customers in 11 states. The utilities estimated that ratepayers would see a 1 percent reduction in rates as a result of efficiencies generated by the merger. The majority of those savings come from personnel reductions. Despite minimal evidence of any benefits from the merger, it was approved by federal and state regulators in 2000.

Given the fluid situation in the energy sector because of changes in rules at the federal and state level, the at best trivial cost savings produced by mergers, and the increasing market concentration that could potentially limit the amount of new generation that comes on line, FERC and state regulatory commissions should oppose any further mergers. At a minimum, the burden of proof should be on utilities to prove that the positive impacts of their mergers outweigh the negative impacts of removing control and authority even further from their customers.

The burden of proof should be on utilities to prove that the positive impacts of their mergers outweigh the negative impacts of removing control and authority even further from their customers.

Rule #2: Maintain the tax exemption for customer-owned utilities

Municipally owned and cooperatively owned utilities have traditionally been given the right to issue tax exempt bonds to finance the construction of distribution, transmission and generation capacity. The use of such bonds lowers their cost of capital and thus reduces their overall cost of operation.[16] Tax exemption is an incentive for customer ownership.

Bonds can also be issued to purchase the lines and poles from the existing investor-owned utility. There has been increasing interest in this option with deregulation. In 1998 and 1999 about 150 communities indicated some interest in buying the poles and wires of their investor-owned electric utilities.[17]

In an era of deregulation, investor-owned utilities argue that tax-exempt bonds would give customer-owned utilities an advantage. The question is whether as a matter of public policy we want to give customer-owned utilities an advantage. Given the evidence that COUs are at least as efficient as IOUs and often a great deal more responsive to their customers, and given the inherently democratic nature of COUs as compared to IOUs, incentives like the use of tax exempt bonds should be continued.

In 1999 the American Public Power Association launched a "Local Control Campaign" to promote The Bond Fairness and Protection Act, a legislative compromise that would allow munis that opt to serve only their residents to continue to issue tax-exempt bonds. Municipal utilities that choose to compete in a deregulated environment would not retain this privilege. The bill would also protect the $75 billion in outstanding tax-exempt bonds currently held by public utilities in over 40 states, regardless of whether these munis choose to participate in a competitive environment.[18]

Rule #3: Encourage place-based energy companies

In the deregulation game large customers get a better deal than smaller customers. The steel mill does better than the local grocery store or bank. The office building does better than the household. To offset the advantages of bigness, small consumers have been working together to combine their buying power to increase their purchasing leverage to get a better price. Often this is done via a third party, a broker or aggregator. The vast majority of aggregators have ownership structures similar to investor-owned utilities. Indeed, often they are closely allied with, or even subsidiaries, of IOUs.

But a number of cities and place-based entities such as chambers of commerce can also act as "energy brokers" or "wireless utilities" and solicit electricity on behalf of their citizens or members that reflect their values and preferences. In this way, consumers can still have a say in how their electricity is generated and where it comes from without actually having an ownership stake in it.

Many such "place-based energy companies" have already been formed:

- The North Central Massachusetts Chamber of Commerce, the Wachusett Area Chamber and the Quabbin Chamber have negotiated a contract that will give their 2,350 member businesses the option to purchase electricity below utility standard offer prices, which are 2.8 cents/kWh to 3.2 cents/kWh in Massachusetts.

- Leaders from 14 municipalities in Bergen and Passaic counties announced the formation of Community Choice NJ. The cooperative effort potentially could benefit 250,000 residents and businesses by aggregating as much as 2.6 billion kilowatt hours of electricity worth more than $300 million per year. At least six other municipalities are considering joining the project.

- Voters in Oneida, New York, have approved a referendum in which the city will aggregate loads. Oneida has a peak load of about 26 MW, including a milk processor, a hospital and city offices that each spend about $500,000 annually on power.

- Leaders of the township of Hampton, Pennsylvania, signed up about half of its 6,400 households who agreed to have their municipal government act as their energy broker. The city negotiated a deal with Allegheny Energy. Residents will save about 6.7% compared to the rates they previously paid to Duquesne Light.[19]

- The Chicago Housing Authority is attempting to aggregate some 40,000 low-income customers that are clients to the authority.

Customer-owned cooperative aggregators also exist, but the circumstances that enable them are rare. For example:

- In late 1998 The National Rural Electric Cooperative Association (NRECA) recognized the 1st Rochdale Cooperative Group, New York City, as its first urban cooperative. 1st Rochdale will serve its member housing cooperative residents in New York City. More than 500,000 New York City families live in and own housing cooperatives.

- The NRECA also welcomed as a new member the California Electric Users Cooperative (CEUC), a cooperative whose 10,000 members are a diverse group of agriculture cooperatives including primarily growers of citrus, dairy, cotton and avocado crops.

Wherever possible states and cities should encourage place-based, customer-owned energy companies.

Rule #4: Make the community the default provider

In an era where electricity customers are being allowed to choose their electricity supplier the question inevitably arises, "What happens if they don't choose?" This is not a hypothetical question because in most states the vast majority of customers aren't choosing a supplier, and they may never want to. By October 2000 only 186,000 customers or 1.8 percent of the more than 10 million eligible electric customers in California had asked to switch suppliers, although a higher percentage of commercial customers had switched.[20]

Because of the inertia and passivity of utility customers, convincing customers to shift to a new supplier can be very costly. Enron reportedly spent $10 million to attract 30,000 residential customers, a $300 per customer cost. Green Mountain Energy Resources managed to acquire a total of 57,100 customers in California and Pennsylvania, but spent $33 million to do so—a cost of $600 per customer.[21] Electricity marketers are losing money signing up customers, but expect to make that money back over the long run as the customers, once signed on, stay with them.

In the era of deregulation, the value of customers is very high. It is a seller's market, which has led to the phenomenon of utilities actually buying customers. As Eugene Coyle, a San Francisco energy-economist, noted in *Local Power News*, GPU, a Morristown, New Jersey-based utility holding company, and Cinergy Corporation, a Cincinnati-based energy company, announced in early 1999 that they will sell their 2.2 million British electricity customers for $300 million while continuing to own and operate Midlands' distribution lines and substations. "This is the first time a regional electricity company's retail supply business has been separated from its distribution business, signaling an emerging phenomenon under electric deregulation in which captive customers are bought and sold as a form of property," says Coyle.[22]

Enron has formally proposed to California and Pennsylvania that customers be auctioned off to the highest bidder. An Ohio legislative committee on electric utility deregulation in 1998 proposed to group consumers who do not seek their own suppliers into Regional Marketing Areas and auction them off to power companies.

In virtually all states, the existing utility has been declared the default utility. This made sense, perhaps, when the existing utility owned both the distribution system and power plants and rates were based on an allocation of costs among these investments. But increasing numbers of local utilities are making money simply by delivering electricity to the customer, not by generating it. And where utilities are permitted to continue owning generating capacity, regulatory commissions are requiring them to construct a "firewall" between their generating and distribution divisions so that they are treated as if there were separate enterprises.

Thus the local distribution utility increasingly serves simply as a transporter of other supplier's electricity. The question of who should be the default utility may therefore better be framed by policymakers as, "which entity would best represent the residents of the area?"

The Massachusetts Model In Massachusetts, due to the remarkable work of then Selectman Matt Patrick of the Town of Falmouth, and Barnstable County Commissioner Rob O'Leary and electricity expert Scott Ridley, cities and towns can purchase electricity on behalf of their residents if the local government votes to do so. The city becomes the default provider. Individuals still have the choice of opting out and choosing their own supplier directly.

This "opt-out" policy differs starkly from the "opt-in" policy most states have adopted, in which individual households and businesses must decide to opt in to a locally managed system. While opt-in aggregation has been successful in Hampton, Pennsylvania, and Oneida, New York, as listed earlier, the experience of private companies like Enron in California, who despite their marketing prowess paid a fortune to sign up only a few thousand customers, suggests that most cities will not find success in an opt-in environment.

In Massachusetts, where local control is favored over absentee control, the rules link authority and responsibility. Advocates of an "opt-out" aggregation system call it Community Choice.

In Massachusetts, where local control is favored over absentee control, the rules link authority and responsibility. Advocates of an "opt-out" aggregation system call it Community Choice.

Cities provide essential services like fire, police, ambulances, trash collection, water and sewer services. They may decide to contract out for these services, in which case the city acts as a broker for its households and businesses. And as Matt Patrick argues, the same should be true with electricity.

In Massachusetts, cities that have become default brokers on behalf of their citizens have begun to negotiate with suppliers not only for lower rates but for a menu of power options. Lexington, Massachusetts asked suppliers to submit bids for five choices: low-price choice, two "green" choices for power from renewable sources like wood chips or trash and from solar, wind or hydro, a choice of energy plus service, and a choice of energy and service plus demand management.[23]

Communities in Massachusetts opting for Community Choice not only have the opportunity to bargain on behalf of their citizens, but they also are entitled to administer the energy efficiency funds collected from their residents.[24] This represents tens of millions of dollars of expenditures that will be decided, not by IOUs that have an economic incentive to wheel as much power as possible over their wires, but by local governments that have a direct interest in shaving off peak loads to improve their load profiles and secure better prices for their power. For example, the Cape Light Compact, a consortium of 20 towns and Dulles and Barnstable counties on Cape Cod and Martha's Vineyard, which represents 185,000 residential and business customers, will receive approximately $25 million in conservation funds over the next five years.

Unlike Commonwealth Electric, which up until now has administered the Cape's energy-efficiency programs, the Compact will not require a 12.9 percent return on investment on the costs of its program.

Municipal aggregation also promises greater investment in renewable electric generation. Susan Munves, conservation coordinator for the city of Santa Monica, envies her counterparts in Massachusetts. California's law, she says, "limits the city to thinking about the environmental impacts of its municipal buildings, and limits consumers to thinking about paying extra to clean up pollution from one home in a half a million homes. With Community Choice we could bring renewables and conservation to the whole community. Rather than having a debate on whether to spend public money to clean up a narrow slice of the city's electricity pollution, we could talk about leveraging community buying power to clean up the whole pie.

In the big picture, cities could dramatically expand the market for wind, solar and other renewable power sources that are now being marginalized because of deregulation."[25]

The principle of community choice has recently been embraced beyond Massachusetts' borders. Ohio's restructuring law, passed in July 1999, included a provision for opt-out municipal aggregation. In a March 7, 2000 referendum, residents of Parma, Ohio, a suburb of Cleveland with 88,000 residents, became the first citizens to take advantage of the new law when they voted overwhelmingly for community choice (16,923-7,123). Of the 136 additional Ohio cities that had community choice measures on the November 7, 2000 ballot, an astounding 132 were approved.[26]

Twelve California cities and counties—including Marin County, San Francisco and Oakland—are eager to follow the lead of Massachusetts and Ohio. Together they have passed a resolution asking the California legislature to amend the state's deregulation law to allow for community choice. Community choice legislation was introduced in the 2000 session, and like the Massachusetts law includes both opt-out municipal aggregation and provisions for communities to use ratepayer fees designated for renewable energy and conservation programs.

In Maine, 11 towns and cities, among them Portland, passed resolutions in favor of community choice plans. The Maine PUC rejected the cities' request and denied a bid from Cumberland County. In New Jersey a strong community choice provision was introduced and then changed dramatically on the last day of the legislative debate.

Rule #5: Encourage customer control of the transmission system

So far we have dealt with the question of who owns and controls the distribution system, that part of the electricity delivery system that reaches into our homes and offices. But an equally important question is who owns and controls the transmission system, those high voltage, bulk carrier lines that move electricity long distances. The rules developed for transmitting electricity will significantly affect the future scale and structure of our electrical system.

Until the mid 1990s, 150 IOUs owned about 78 percent of the transmission lines. Sixty G&T cooperatives owned 8 percent and munis and state and federal public-power authorities owned about 14 percent. Regional power pools coordinated the modest flow of electricity between jurisdictions. The power pools were made up of representatives of the individual transmission-owning utilities.

The rules developed for transmitting electricity will significantly affect the future scale and structure of our electrical system.

The Energy Policy Act of 1992 directed the Federal Energy Regulatory Commission (FERC) to develop rules giving all electricity suppliers equal access to the 400,000 miles of transmission lines that crisscross the country. A key to accomplishing this has been to develop a governance and operational structure that stops utilities from self-dealing: that is, from favoring the transmission of their own generating capacity over some other generator.

To date FERC has given existing transmission authorities and state agencies wide latitude in developing new structures.

No consistent organizational structure or configuration has emerged for regional transmission systems. Originally FERC wanted regions to form regional transmission groups (RTGs), which would be voluntary groupings of utilities covering the same geographic areas as the present coordinating regions. Later, FERC urged regions to consider forming independent system operators (ISOs), which would be trustees for the owners. The ISO would control the system and develop transmission pricing rates and undertake strategic planning for the transmission system. As of late 1999, six ISOs have been approved by FERC: California; Midwest; New England;

New York; Pennsylvania-New Jersey-Maryland (PJM); and Texas. Five more were under formation.[27]

While in theory ISOs are governed by individuals with no financial stake in any market participants, this isn't always the case. FERC has not made any enforceable rules against utility holding company representatives serving on an ISO's board of directors. Furthermore, in California, "the state itself not only has no control over the operations of California's independent system operator (Cal-ISO); it cannot even require the Cal-ISO or the California Power Exchange to turn their records over to the state attorney general to investigate possible collusion."[28]

The reason for the creation of the ISOs has been as a vehicle to ensure nondiscriminatory access to the transmission systems. Utilities often give their own power plants priority access to their own transmission lines. FERC itself has noted the difficulties it has encountered in eliminating self-dealing and discrimination against independent power producers (IPPs). FERC has observed that the increase in the number of market participants and transactions in wholesale markets has made discriminatory behavior more difficult to detect. The Commission has also said that the functional unbundling of utility activities—which was required under Order 888 of 1996—has not produced sufficient separation between operating the transmission system and marketing and selling power, and that this lack of separation contributes to discriminatory behavior.[29]

Several of the ISOs have developed an advisory board of stakeholders, such as residential consumers, environmental groups, power marketers, large industrial customers, IOUs and COUs.

But many have no small consumer representative in decisionmaking positions. The New England ISO, for example, has no government official or consumer advocate on its Board. The public has representation only on an advisory panel.

In yet another effort to make the transmission lines truly common carriers, FERC issued another order in late 1999—Order 2000—calling for the voluntary creation of regional transmission organizations (RTOs). RTOs are to be completely independent from power production and sales, so as to sever the economic incentives between power marketing and control of the transmission system. RTOs can either be based on the ISO model or on the transmission company model, in which the so-called transco is an independent, profit-making transmission company that owns the transmission facilities. Under Order 2000, utilities that were not members of an ISO were required to submit plans to join an RTO by October 2000; utilities that were members of an existing regional organization were required to submit their plans to join an RTO by January 2001.

Since the governance of transmission lines will be separated from their actual operation, and since the operation of the transmission system directly affects the operations of the distribution systems, then the boards of directors should be directly elected by their customers.

Order 2000 spells out three criteria that RTOs must meet to demonstrate their independence: "(1) the RTO, its employees, and any non-stakeholder director must not have any financial interest in any market participants, (2) the RTO must have a decision-making process independent of control by any market participant, and (3) the RTO must have exclusive authority under Section 205 of the Federal Power Act to file changes to its transmission tariff."[30]

Since the governance of transmission lines will be separated from their actual operation, and since the operation of the transmission system directly affects the operations of the distribution systems, then the boards of directors should be directly elected by their customers. The voting structure could be varied. Some regions might want to have representatives elected from districts. Others might want representation by different sectors (industry, residential, low-income, etc.).

In early 2001, the California legislature began debating a bill introduced by the Senate President pro tem John Burton and supported by Governor Gray that would have the state purchase the 32,000 miles of transmission lines from the state's three private utilities. The buyout would be the quid pro quo for the state's multibillion dollar bailout of the utility companies, which were near bankruptcy because of the runup in electricity prices in late 2000. The purchase would give the state ownership of 60 percent of the high voltage transmission lines in the state. The other 40 percent are currently owned by municipal utilities, cooperatives and federal agencies. Burton explained the proposal in this way, "What we're trying to do here is give the state some influence and control over its own destiny."

Rule #6: Minimize rather than maximize the geographical area served by the regional transmission agencies

FERC insists, in its ISO Principle 3, that these structures should be "as large as possible." Thus for example, the Midwest ISO would cover an 8-state region with 32,000 miles of transmission lines, serving almost 9 million customers, with 63,000 MW of generating capacity.

Some, like engineer Kiah Harris, argue that since the country is divided into three separate interconnections systems (West, Texas and East), there should be three commensurately large governance structures. Greater transmission flows mean greater impacts on more remote transmission lines, which requires a larger area of control.

Order 2000 does not establish an appropriate size for an RTO. It does acknowledge that one size does not fit all regions, so different sizes and configurations are likely. FERC encourages RTOs to maximize their size, as expressed explicitly in its early ISO Principle 3.

Although the argument in favor of ever-larger transmission governance structures makes sense from an electrical engineering perspective, the argument against it is that it removes decisionmakers from those who feel the impact of their decisions. Moreover, it creates a decisionmaking structure which, because of its jurisdiction, will make decisions that favor long-distance transportation of electricity when we may be entering an era in which, because of the rise of decentralized power and a better understanding of how dispersed power benefits the overall system, the presumption should be to favor minimizing the distance traveled by electricity.

Rule #7: Restrict the right of eminent domain

In the 1920s electric utilities were given governmental authority to seize private property to build high-voltage transmission lines. This authority is called eminent domain. The public good of getting electricity from point A to point B was considered greater than the rights of private property owners.

Although the argument in favor of ever-larger transmission governance structures makes sense from an electrical engineering perspective, the argument against it is that it removes decisionmakers from those who feel the impact of their decisions.

There has always been considerable opposition to these high-voltage transmission lines, but opposition has grown as the voltage levels have risen to 345 kV and 745 kV. The opposition arises out of aesthetic and health concerns in addition to the feelings of a loss of control.

As a result of public pressure, California recently dropped the minimum voltage requirement for seeking a construction permit from 200 kV to 50 kV to give the public more opportunity to intervene. Pennsylvania and Texas have broadened their application and public-hearing rules.

Many states already have sufficient transmission capacity to serve local needs. If local communities are asked to permit new transmission lines to cross their territories for the benefit of distant buyers and sellers, they may be reluctant to oblige. And what if the transmission system is operated by an enterprise with no history of local service? As one regulatory official put it, "Without the moral authority that local power companies have always had, it's going to become a lot harder to build anything."[31]

The decision on siting transmission systems has traditionally rested in the hands of states and local governments, but if in the future public opposition stymies expansion, the federal government may intervene and pre-empt local and state authority. As Ashley Brown, former chair of the Electricity Committee of the National Association of Regulatory Utility Commissioners and former Commissioner of the Ohio PUC told a Congressional committee, "FERC can mandate under section 211 that a utility provide access to any eligible party that seeks it, but if the party from whom access is sought can show to FERC's satisfaction that it cannot provide that access with existing facilities, it then has a good faith obligation to go to the state siting authority and get that line sited."[32]

Kiah Harris notes that current ISO proposals do not include the requirement that the ISO build transmission facilities. Proposed projects will be reviewed by the ISO, but the responsibility of construction is up to the sponsor of the project. "Since the projects are going to be built by private companies, this implies that there will be no ability to acquire right of way except through commercial means."[33]

The argument in favor of allowing private property to be seized to make way for high-voltage transmission lines has been undermined by the rise of small-scale power plants. Increasingly states may require utilities and regulatory commissions to estimate the comparative cost of transmission versus dispersed power (or storage or efficiency). In

Wisconsin such "targeted area planning" is already required. As this occurs the public will gain a better sense about the tradeoffs involved. If the cost increase is trivial (e.g. 0.2 cents per kWh or less), the decision might be to forego the installation of such lines.

Before any private property is seized to construct a transmission or distribution line, an analysis should be undertaken to ascertain what the comparative cost would be of improving energy efficiency, or energy storage or dispersed electric generation instead. The burden of proof should be on those proposing the new high-voltage transmission line.

RULES FOR ASSUMING AUTHORITY

1. Declare a moratorium on large energy mergers at the state and federal level
2. Maintain the tax exemption for customer-owned utilities
3. Encourage place-based energy companies
4. Making the community the default providee
5. Encourage customer ownership of the transmission system
6. Minimize rather than maximize the geographical area served by the regional transmission agencies
7. Restrict the right of eminent domain

CHAPTER FIVE

Accepting Responsibility: Protecting People and the Environment

The electricity system ties us together into a web of mutual responsibility. At the block level we share a transformer (the device that lowers and raises voltage) and thus become an electrical unit. The behavior of an individual homeowner (e.g. turning on power equipment) can affect the electricity levels and quality in other homes on the block. At the global level we share a biosphere. Americans consume up to 100 times as much electricity as the average person on the plant, yet the emissions from our power plants are affecting the weather patterns of people half a world away. Sulfur emissions from Midwest power plants falls as acid rain on New England.

For a growing number of us, continued access to electricity is essential for survival. Our furnaces depend on tiny pulses of electricity to turn on. Millions of us reside in desert-like locations inhabitable during the summer months only because of electric air conditioning. Electricity is the underpinning of an information economy.

The restructuring of the electricity system forces us to confront the issue of responsibility head-on.

In a regulated system, utilities had the "obligation to serve." In a deregulated system, they do not. Some believe that the California crisis in early 2001 was a result of a botched deregulation system in which the distribution utilities still had an obligation to serve and not to raise prices, while the independent suppliers of electricity no longer had either. But if California had instead allowed utilities to pass through the increased charges, the enormous increase in rates would have cut off many poor households from electricity. In a competitive system, how do we guarantee universal access to electricity?

In a regulated system, state agencies required utilities to encourage environmentally benign sources of power. These often had a slightly higher price, but had nonquantifiable benefits that outweighed the modest price increase. In a competitive system, how do we "internalize" the environmental costs of power production?

In a regulated system, power plants became a significant source of tax revenue for local communities. In a competitive system, an in-state power plant that must pay such taxes will be at a cost disadvantage compared with an out-of-state electricity provider that does not. If we eliminate these taxes, who will make up the financial shortfall for our schools and libraries and public services?

Virtually every state is tackling these questions. No uniform answers have emerged. In this chapter we offer rules that make us responsible, to this generation and to the next generation.

Protecting the Poor

Rule #1: Low income households must have access to electricity

In return for their monopoly status, electric utilities have an obligation to serve all members of the community. Over the years, a number of states have elaborated rules to specifically protect low-income households' access to electricity. These have taken four forms:

a) a ban on power disconnection in winter
b) energy assistance payments and the absorption of delinquent bills by other utility customers

c) lifeline rates and discounts
d) low-income energy conservation (weatherization) programs

Some states have formally required providers and utilities to continue and expand programs targeted to low-income households under a restructured system.

Some states have formally required providers and utilities to continue and expand programs targeted to low-income households under a restructured system.

- Massachusetts requires that distribution companies continue programs "comparable to the low-income discount rate in effect prior to March 1, 1998." Eligibility may extend to 175 percent of federal poverty guidelines.[1] Included in the energy conservation program is a permanent set aside for low-income energy efficiency investments of 0.25 mills per kWh or 20 percent of each utility's residential conservation program. Coordination is carried out by the local Weatherization Assistance Program agencies.

- California's electric restructuring law, AB 1890 Section 1(d), states "It is the further intent of the Legislature to continue to fund low-income ratepayer assistance programs..." California's CARE program provides a 15 percent discount on gas, and electric and monthly customer charges to households with incomes at or below 150 percent of federal poverty guidelines.[2]

- In Pennsylvania, the Consumer Choice Act, effective January 1, 1997 required programs to allow low-income customers to maintain electric service. The

Fig. 12. Annual State-Level Low-Income Funding Before and After Restructuring

State	Low-income funding prior to restructuring ($million)	Low-income funding after restructuring ($million)
California	66.7	81.0
Connecticut	2.6	2.6
Illinois	0.5	26.0
Maine	6.1	6.1
Massachusetts	37.6	46.0
Montana	2.4	2.4
New Hampshire	0.5	TBD
New York	NA	TBD
Oklahoma	2.0	2.0
Pennsylvania	82.0	82.0
Rhode Island	2.4	2.4

TBD = To be determined

NA = Not available

Source: Energy Programs Consortium

act requires distribution utilities to rely on community-based organizations for the delivery of these programs where that is appropriate.

- New Hampshire's electric restructuring legislation calls for "programs and mechanisms that enable residential customers with low-incomes to manage and afford essential electricity requirements..." Maine's requires a minimum of 0.5 percent of distribution electric utility revenues be spent for low-income programs.

Overall, funding for low-income households held firm or modestly increased after restructuring, although in many cases the before and after comparison is misleading because utilities decreased their spending on low income-households significantly between 1993 and 1998.

Protecting the Environment

Electric generation is the single largest source of air pollution. According to the U.S. Environmental Protection Agency and Energy Information Administration, electricity generation is responsible for 67 percent of sulfur dioxide, 25 percent of nitrogen oxide, 36 percent of carbon dioxide and 33 percent of mercury emissions.[3]

Electric power plants are major contributors of greenhouse gases. In 1997 they accounted for 36% of total U.S. greenhouse gas emissions and over 8% of the world's emissions. Also, according to the Renewable Energy Policy Project, electric power plants emitted 1 billion pounds of toxins in 1998, more than the chemical, paper, plastics and refining industries combined.[4]

Power customers have been clear in their desire for clean electricity and have even expressed a willingness to pay a premium to get it. In a paper analyzing over 700 polls conducted between 1972 and 1996, Barbara Farhar shows that consumers have an overwhelming preference for green energy systems. More importantly, "approximately 56% to 80% of respondents to recent national surveys say they would pay a premium for environmental protection of renewable energy."[5] Dozens of utilities currently offer their customers the option of choosing to buy "clean" energy to meet a part or all of their needs.

Rule #2: Green marketing: favor green citizenship over green consumerism

Green-pricing programs, in which customers are asked to voluntarily pay a premium for varying amounts of electricity generated by renewable fuels, are sweeping the country.

Green-pricing programs encourage environmentally oriented people to put their money where their mouth is. They encourage electricity marketers to develop national educational campaigns that promote clean energy. This has indeed occurred. In fact, the largest single reason for residential customers to purchase electricity from an independent supplier is to buy green energy. In California, 90 percent of households that have switched suppliers have voted for green power with their electricity dollars.[6] Companies like Patagonia and Toyota and cities like Santa Monica have voted to purchase electricity that is partially or fully generated by renewable energy for their internal use.

However, while welcome, green consumerism suffers serious limitations.

Green-pricing programs impose a very stiff premium on consumers who want to be responsible, and in the aggregate, generate a relatively small amount of green-demand. In some cases, consumers are buying power from existing renewable energy projects. In effect, renewable energy producers, like existing geothermal-fueled electricity or wind power facilities, are reselling their electricity at higher prices. Supporters of green pricing say that this is a short-term effect until the current capacity is soaked up. Others argue it will take a long time before that point is reached.

Utility expert Nancy Rader notes, "If every customer in eleven western states had the choice of purchasing renewable electricity, 18 percent of those consumers would have to purchase 100% non large hydro renewable product in order to subscribe the existing amount of renewables that are either included in utility ratebases or are under long-term contract to a utility with costs passed through to ratepayers."[7]

Yet the highest participation rate for a green pricing program thus far is the meager 3.4 percent enrollment achieved by Michigan municipal utility Traverse City Light and Power's wind energy program. And in this case the circumstances were ideal. It is a small, isolated town. A state grant covered 8 percent of the costs. The wind turbine is two miles from city limits so people can see it. There was a major media campaign and a targeted direct mail effort. And the price premium was only 23 percent, or 1.58 cents per kWh.

Most green-pricing programs force participants to pay 30 percent or more over the average retail rate. For businesses the price hike could be more than 60 percent.[8]

Green pricing requires a few customers to pay a substantial premium for relatively little power.[9] A much better way for consumers to increase the supply of renewable energy is to exercise "green citizenship." If a significant majority of the customers of a given utility vote for green energy, the utility can purchase a larger amount of renewables and spread the costs over its entire customer base. Often 10 times the amount of green electricity can be purchased at a fraction of the cost for an individual household. To date only one utility of which we are aware, the customer-owned Salem Electric Cooperative in Oregon, has adopted this strategy.

Green pricing requires a few customers to pay a substantial premium for relatively little power. A much better way for consumers to increase the supply of renewable energy is to exercise "green citizenship."

Green citizenship, in short, not only dramatically reduces the premium paid for renewable power by spreading the costs out over all customers, but also dramatically increases the amount of green power produced.

COUs, like the Salem Electric Cooperative, are more likely to enact green citizenship programs because they tend to be more responsive to the preferences of their customers. But where there is utility inaction, or where the electricity landscape is dominated by IOUs, there are policies that states can enact that embrace green citizenship and remove the onus of supporting renewables from the individual ratepayer. The two most popular strategies that restructuring states have adopted are the renewable portfolio standard (RPS) and a public benefits fund for renewables.

Rule #3: Set an increasing quota for environmentally benign electricity

As of November, 2000, eight states had adopted a renewable portfolio standard (RPS).[10] This is a performance standard requiring that a certain percentage of electricity sold into a given market by each provider be derived from renewable fuels. Estimates indicate that these eight portfolio standards will stimulate markets for approximately 3,800 MW of new renewables and provide support for about 3,600 MW of existing renewables capacity.[11]

RPSs vary significantly by state. Nevada's is the only one that gives preference to in-state suppliers. The largest quantities of renewable energy required under an RPS are in Connecticut, Massachusetts, New Jersey and Texas. Oil-rich Texas, which does not have a history as a strong supporter of renewable technologies, surprisingly will have the largest impact. The Texas law requires 2,880 total MW of renewables-based electricity by 2009, with 2,000 of those MW required to be from new capacity—more than half the estimated new capacity for all RPSs combined.

While Connecticut's and Massachusetts's RPSs do call for significant increases in renewable generation, they are not guaranteed.[12] In Connecticut, the Public Utilities Commission voted to exempt default service providers, that is, the original utilities, from the RPS. Because so few customers have switched suppliers, this would in effect render the RPS meaningless. The Connecticut Consumer Council has appealed this ruling to the State Superior Court. The Massachusetts law is vaguely written, and it is unclear whether the default utilities in that state will be forced to comply with the RPS.

Texas is the only state that has included renewable energy credit trading as part of its RPS. Five other states (Connecticut, Massachusetts, New Jersey, Nevada and Wisconsin) are considering credit trading. Credits allow flexibility in meeting the goals of the RPS. They encourage suppliers to build renewable capacity not only to meet the requirements of the RPS but also to reap additional revenues by selling excess credits to suppliers who do not meet the RPS requirements with their own capacity.

All states that have an RPS, except Maine, require that electricity suppliers must develop new sources of renewable energy to meet the standard. Thus Maine is in the ironic situation that a strict adherence to its RPS could mean less electricity coming from renewable energy in the future. Its RPS is by far the highest in the nation, requiring that 30 percent of their electricity come from renewable energy. But in 1998 nearly 50 percent of Maine's electricity already was derived from wood or hydro.

Public benefits funds for renewables can be enacted even by states not going through the deregulation process. Iowa and Minnesota, for example, have required their IOUs to purchase specific amounts of electricity. Minnesota's mandate, passed in 1994, requires its major IOU to purchase up to 950 MW of renewable electricity, 825 MW from wind power and an additional 125 MW from biomass-fueled power plants.

Rule #4: Establish a public benefits fund for renewables and high efficiency distributed power plants

Most states have create a direct funding mechanism for renewables. Most commonly, public benefits funding is provided from fees placed on electricity customers, based on how much energy they consume. Such a fee is called a systems benefit charge (SBC), and is somewhat analogous to the fees tacked on to long distance telephone calls to fund universal telephone service.

At least fourteen states have embraced some sort of public benefits fund that covers renewables—twelve of which are SBCs. Other states have public benefits funds that are used to pay for energy efficiency programs, research and development, universal service, and other low-income protections. Five states (Connecticut, Massachusetts, New Jersey, Pennsylvania and Wisconsin) have enacted both an RPS and an SBC.[13] Wisconsin has done so without restructuring its industry.

If structured correctly, Rules 3 and 4 could work in tandem to jumpstart renewables development. For example, state policymakers could prevent projects that receive SBC funds from selling power to RPS markets or could focus SBC funds on renewable technologies and markets that are unlikely to be encouraged by the RPS. If precautions such as these are not taken, interactions between the two policies can affect their aggregate impact.[14]

A continued commitment to research and development (R&D) funding is essential to the development of state-of-the-art renewables and efficiency technologies. As a rule, new technologies for generating and delivering electricity take a lighter toll on the environment than old ones. Unfortunately, in the uncertain environment surrounding restructuring, most utilities have scaled back their R&D efforts. According to the U.S. General Accounting Office, R&D spending by utilities dropped by one-third between 1993 and 1996.[15]

Of all the states with annual funds for renewables, California's annual commitment of $135 million is by far the highest in the nation. This funding, however, is guaranteed only for four years, the shortest duration of any state but Montana. California's per capita spending also falls well short of that of Connecticut and Massachusetts.[16]

Fig. 13. Renewable Energy Funding by State with Duration ($ per capita)

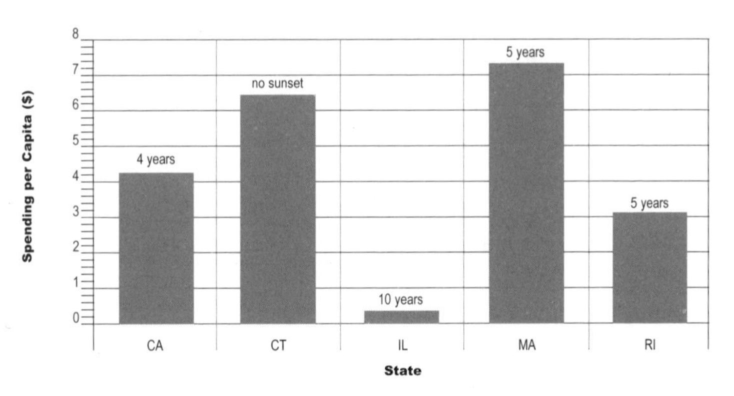

By 2010, the 12 existing state SBCs will have collected a total of $1.7 billion to support renewable energy.[17]

Rule #5: Invest in energy efficiency at levels at least as high as expenditures in the peak year (usually 1994)
The most cost-effective way of reducing pollution is by improving energy efficiency. Improving efficiency is even more important in our computer-driven economy. At present projections, Americans will use 27% more energy in 2020 than in 1998, requiring the construction of 1,000 new power plants.[18]

Efficiency makes sense. But there are at least three significant obstacles to improving efficiency even when the cost-effectiveness is high.

First, most households and business owners will invest in efficiency only if they see a six-month to two-year payback.

Second, households and businesses are reluctant to make long-term investments that will accrue to the advantage of the next building or home occupant, not themselves.

Third, many decisions about the level of efficiency of lights or appliances are made, not by those who will pay the electric bill, but by landlords or developers of commercial office space that favor minimum initial costs, rather than minimum operating costs.

To overcome these obstacles, state legislatures and regulatory commissions in the 1980s and 1990s required utilities to finance energy-efficiency programs. Collective spending on these programs peaked at about $2.4 billion in 1994.

Although some of the money was wasted, and utility efficiency programs have been subject to persuasive critiques, the increased spending does appear to have resulted in dramatic energy savings. Between 1989 and 1994, when electric utility expenditures on energy efficiency programs more than tripled, energy savings also tripled. In 1994 U.S. electric utilities spent about 1.5 percent of their total revenues on such programs, reducing annual sales by 1.9 percent and cutting peak demand by 7.3 percent.[19] Efficiency expert Eric Hirst of Oak Ridge National Laboratory concluded, "annual savings equiva-

Fig. 14. Utility Efficiency Spending as Percent of Revenue: 1993 to 1998[23]

State	Spending as % of Revenues 1993	Spending as % of Revenues 1998
Connecticut	1.79	0.96
Maine	1.27	1.30
Massachusetts	2.55	1.91
New Hampshire	0.33	0.41
Rhode Island	1.98	1.70
New Jersey	0.29	1.34
New York	1.66	0.66
Pennsylvania	0.16	0.01
Illinois	0.02	0.06
Maryland	1.83	0.80
Virginia	0.19	0.00
Texas	0.26	0.17
Arizona	0.24	0.11
Montana	1.61	0.72
New Mexico	0.04	0.11
California	1.40	0.75
U.S. Total	**0.83**	**0.42**

lent to 1 percent of system consumption were being achieved by companies that had in no sense tested the limits of their capacity."[20] Eto et al. examined the 40 largest utilities' commercial energy-efficiency programs and found them to be, on the whole, highly cost-effective.[21]

In a competitive marketplace efficiency investments will shrink because no utility will be willing to raise its prices in the short term to effect long-term savings that may accrue to customers, not shareholders. That thesis has already been proved in practice. The World Wildlife Fund and Environmental Working Group found that in 1997 utilities spent only $894 million on conservation programs, less than 40 percent of the $2.4 billion they spent in 1992.[22] Actual spending on energy efficiency in 1997 was less than one quarter of one percent of revenues.

In a competitive marketplace efficiency investments will shrink because no utility will be willing to raise its prices in the short term to effect long-term savings that may accrue to customers, not shareholders.

States involved in restructuring have reacted to decreasing IOU efficiency commitments by imposing a "wires charge" on electricity, a non-bypassable sales tax on all electricity purchased in the state. The money goes into a public benefits fund, part of which is used to invest in energy efficiency.

The strongest efficiency funding is in Connecticut. Its annual per capita spending is second only to Massachusetts, but Connecticut's statute contains no sunset date and nearly doubles the funding levels of 1994. Illinois and Rhode Island have very weak efficiency programs, where per capita spending is less than $1 and efficiency expenditures as a percentage of 1994 funding levels decreased by more than two thirds. But when Maryland passed its restructuring legislation in March, 2000, it won the shameful distinction of becoming the first state with a previous commitment to energy efficiency to completely abandon that commitment in a competitive market.[24]

Fig. 15. Energy Efficiency Spending by State Before and After Restructuring
($ per capita)

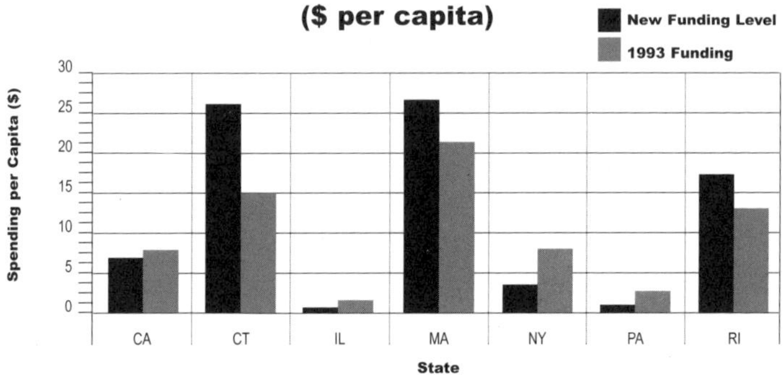

Rule #6: Enact a pollution portfolio standard

A pollution portfolio standard (PPS) is an excellent companion policy to a renewable portfolio standard. The RPS ratchets up the quantity of electricity generated from renewable, decentralized fuels. A pollution portfolio standard ratchets down the amount of pollution generated from nonrenewable, centralizing fuels.

A key reason for a PPS is the resurgence of coal-fired power plants and, especially, the potential for enormous increases of electricity from older, dirtier coal plants. "For at least the next decade, the most important environmental variable for North American electricity is the fate of more than 300,000 MW of underutilized coal-fired generation," writes Ralph Cavanaugh.[25] One reason is that older coal-fired power plants were exempted from the emission-performance standards of the Clean Air Act in 1970 and again in 1977 on the theory that these older plants would be retired in 20 to 30 years. But they continue to operate, and in a deregulated environment many are ratcheting up their annual output.

In the four years preceding 1992's EPAct, the nation's fleet of coal plants increased generation at an annual rate of about 2%; in the six years after its passage, through 1998, generation grew by 15.8%, as the plants boosted operations from 60 to 67 percent of capacity.[26] Electricity demand over these years only increased by a total of 11.6%, meaning that electricity from dirty coal plants is meeting an ever greater portion of our demand.[27]

This increase in coal-fired generation has corresponded with higher levels of pollution. Nitrogen oxide emissions from this extra generation equaled 755,000 tons in 1998, the same amount of smog-forming pollution emitted each year by nearly 37 million cars. In addition, the increase in power generation from these plants was responsible for 298 million tons of carbon dioxide in 1998, an amount equal to the carbon dioxide emitted per year by nearly 44 million cars.[28]

Environmentalists are demanding that coal-fired power plants that are revved up to meet the new market demand must also meet the new emission performance standards. Otherwise dirty coal-fired electricity will compete unfairly against clean coal plants and other cleaner energy producers.

Despite the increased workload of old coal-fired plants, they still have plenty of additional excess capacity. Generation from coal power plants could potentially increase another 20 percent, assuming that plants will run at no more than 85 percent of their maximum capacity.[29]

Environmentalists are demanding that coal-fired power plants that are revved up to meet the new market demand must also meet the new emission performance standards. Otherwise dirty coal-fired electricity will compete unfairly against clean coal plants and other cleaner energy producers. On average, older coal-fired plants that are exempt from the Clean Air Act produce electricity at a price about one cent per kilowatt hour lower than new plants.[30]

A PPS would require that suppliers not exceed maximum pollution emission levels. It could also require suppliers to gradually improve upon their emission levels, just as an RPS requires suppliers to gradually increase the percentage of their renewable-derived energy.

Connecticut and Massachusetts have moved in this direction by directing their environmental regulators to develop emission performance standards for retail supplier portfolios. Such standards could also embrace trading schemes, as the Texas RPS has. The EPA has proposed a nitrogen oxides trading scheme for Eastern states.[31]

To better inform consumers making decisions about electricity suppliers, a number of states are requiring that utilities disclose the amount of pollution generated by each marketer. Illinois, for example, requires that electric bills beginning January 1999 list the percentage of electricity supplied by each fuel source (e.g. biomass, coal, hydro, gas, nuclear, solar). Emissions data must be provided quarterly by utilities, including nuclear waste, carbon dioxide, nitrous oxides and sulfur dioxide emissions. Massachusetts and Connecticut also require the disclosure of emissions data in addition to fuel mix. On a regional level, New England states and the Western Governors' Association are exploring a uniform regional disclosure approach.[32] Thus it should be possible to use this data to monitor an emission standard.

Rule #7: Substitute pollution taxes for property taxes

In the 1980s and 1990s several state regulatory commissions quantified the environmental costs of electricity generation for use in comparing power plant costs. Their estimates varied widely, but all agreed that there is a substantial environmental cost not included in the price customers pay for electric power. These "external" costs are borne by society as a whole in the form of environmental degradation and through medical costs as a result of pollution from power plants.

These external costs can be internalized by imposing a pollution tax on electricity. Pollution taxes and environmental tax shifting have been discussed for many years. The argument behind them is unassailable. Today we tend to tax those things we would like to encourage, like work and property and income, while we undertax or don't tax at all those things we would like to discourage, like pollution and waste. Why not change the dynamic by increasing taxes on waste and decreasing taxes on work?

Pollution taxes and environmental tax shifting have been discussed for many years. The restructuring debate offers a window of opportunity for tax shift advocates.

This argument, while persuasive, has not resulted in environmental tax shifting in the United States. One reason is that in an era of tax cutting it is virtually impossible to generate serious discussion about imposing a new tax, even if the result is revenue neutral (i.e. the amount of revenue generated from the new tax would be offset by an equal reduction in an existing tax).

The restructuring debate offers a window of opportunity for tax shift advocates. Today utility taxes make up a significant source of revenue for communities that host power plants. In a monopoly system these taxes are easily passed through by utilities to customers. But when a customer can choose an out-of-state supplier, then a local tax on electricity could become a competitive burden. Thus, virtually all states that are moving toward utility restructuring are embracing a "tax shift" in this area, substituting one tax for another. Illinois, for example, has replaced its utility taxes with a ten-tiered tax based on electricity use. New Jersey has eliminated most of its taxes on electricity and substituted a flat 6 percent sales tax on electricity.

Nationwide more than $15 billion a year is generated from electricity-related fees and taxes. That comes to about 0.4 cents per kWh.[33] To raise the same amount of revenue would require a nationwide carbon tax of $75 per ton. Interestingly, this would raise the price of coal-fired electricity by almost 2 cents per kWh, surpassing the existing federal tax credit to renewable electricity producers.

An electricity-related environmental tax-shift strategy must be state specific. Convincing coal-rich Illinois or West Virginia to impose a carbon tax might be difficult but the tax imposed per unit of carbon emitted would be very small because of the large amount of carbon emitted. Convincing the Pacific Northwest to adopt a tax shift might be easy politically, but because 85 percent or so of its electricity comes from hydropower, the tax per unit of carbon would have to be extremely high.[34]

This type of pollution tax can be structured to be revenue neutral; that is, the same amount of revenue is generated after the tax as before the tax. However, it is also possible to raise the pollution tax to pay for public benefits programs that also protect the environment and safeguard low-income households. If these programs were included, the total tax might rise by 25 percent.

Protecting the ratepayer: Who should pay for the mistakes of the past?

> "Stranded costs are a brooding omnipresence. Like the ghost of King Hamlet, these costs demand that historic accounts be squared so that the poor souls now weighted down by them can pass from their present uncomfortable limbo to a more agreeable place."[35]
>
> *Irwin Stelzer, director of Regulatory Policy Studies at the American Enterprise Institute*

By the mid 1990s, it had become cheaper to build a new high-efficiency natural gas-fired plant than to operate many existing nuclear power plants. If power generation were to become a competitive market, this meant that billions of dollars of existing utility assets could suddenly become liabilities, so-called "stranded costs."[36] The question of who should pay for these losses—shareholders or customers—has been a highly contentious issue, especially in those 20 or so states where utilities bet heavily on nuclear power or where regulatory commissions or legislatures established a very high long-term price for independently purchased power.[37]

IOUs argue that the customers should continue to pay 100 percent of the uneconomic costs. They maintain that a "regulatory compact" exists, that states had allowed the investments in the first place and cannot, after the fact, change the rules, especially when the utilities kept their end of the bargain by providing reliable, universal service.

The regulatory compact argument has not been accepted by the courts. In the 1980s, regulatory commissions regularly denied utilities the right to earn a return on investments already made in nuclear power plants and courts upheld their actions. "An examination of the origins and content of the regulatory compact finds little basis for the claim that utilities are always entitled to cost recovery and a return on their past investments," concludes Dr. Kenneth Rose, senior institute economist for the National Regulatory Research Institute, an organization funded by the National Association of Regulatory Utility Commissioners (NARUC).[38]

Most states that have gone through the deregulation process have allowed utilities to impose 100 percent of their stranded costs on their customers, although there has been a great deal of variation in their approaches.[39]

Two states have made efforts to deny utilities full recovery of their stranded costs. In 1998 New Mexico's Public Regulation Commission denied any stranded cost recovery to Public Service of New Mexico because, "Under New Mexico law, the utility duty to render efficient service precludes the recovery of stranded costs, which are, by definition, a measurement of inefficiency. Having never had an entitlement to recover inefficient costs, a utility may not claim an unconstitutional taking when not allowed recovery of inefficient costs."[40] But in 1999 the state legislature passed restructuring legislation that overrode the PRC's decision. Still, the new law only guarantees 50 percent recovery.

New Hampshire's regulatory commissioners allowed recovery of stranded costs only if the utility's rates were at or below the regional electric average electricity rates. If the local utility had rates higher than a neighboring utility one could suspect management error and therefore investor liability. Full recovery would be permitted only in those cases where the utility management's discretion over resource acquisition was reduced or eliminated by government mandates.[41] Based on this formula Public Service of New Hampshire (PSNH) was allowed only 60 percent of its stranded costs.

"Under New Mexico law, the utility duty to render efficient service precludes the recovery of stranded costs, which are, by definition, a measurement of inefficiency."

PSNH took the commission to court, and in April, 1999, a U.S. District Court judge froze New Hampshire restructuring activity and ordered the PUC to reissue its stranded cost estimate. In June, the two sides reached an agreement that will allow the utility to recover 85 percent of its total of $2.3 billion in stranded costs. PSNH will be allowed to securitize $725 million of that total. The utility is also required to divest its generation assets by July 2001. Profits from the sale of these assets will be applied to the portion of the stranded costs for which PSNH is responsible.[42]

Ironically, stranded costs are based on the comparative costs of existing power plants to new power plants. In the mid 1990s, new power plants, fueled by natural gas, were cheaper than many existing power plants. But when natural gas prices rose sharply in 2000, and as increasing customer demand finally soaked up the surplus of electricity generating capacity built in the 1980s, electricity prices rose and the value of so-called "stranded" assets rose as well. Sometimes this led to bizarre situations. California, for example, forced its electricity customers to pay over ten billion dollars to a handful of utilities to pay for noncompetitive utility assets. The utilities sold the assets, sometimes at depressed prices. Four years later, those assets are very valuable, electricity costs are soaring, and these same utilities that negotiated the original agreement are petitioning the courts to order the regulatory agency to increase customer rates again, this time to pay for their operating losses.

There is no legal requirement that customers bear 100 percent of the losses from past utility investments. Neither is there any standard that would help us to allocate the costs between shareholders and customers. Adam D. Thierer of the Heritage Foundation offers a simple standard: "If a utility can show that it made an investment only at the insistence of regulators and that it actively had resisted the action but was forced to move forward anyway, then it has a better case for compensation."[43]

Rule #8: Stranded costs should be paid by the party primarily responsible for causing those costs

According to Resource Data International, two categories of stranded costs account for more than half of the total: $86 billion in nuclear plants and $42 billion in purchased power contracts.[44] Using Thierer's guideline, the utility's owners deserve to bear the brunt of the excess nuclear costs while its customers deserve to bear the brunt of the excess independent power costs.[45]

Nuclear Power

In the late 1970s and early 1980s, when utilities requested permission to build nuclear plants, environmentalists vigorously argued that nuclear plants were expensive, that electric demand was no longer increasing as it had in the past, that improving efficiency was cheaper than building gigantic new power plants, and that the lack of acceptable radioactive waste storage facilities made nuclear power's future uncertain.

Utilities spent hundreds of millions of dollars to convince state regulatory commissions and legislatures that nuclear power was a low-cost and viable source of electricity. In most states these efforts were successful.

Now, 20 years later, the evidence is in. Opponents of nuclear power were right. Today, nearly half of all utility generation investments are accounted for by nuclear generation, even though nuclear plants generate only a little over 20 percent of the nation's electricity. On the open market fossil-fueled plants have been selling at a premium while nuclear power plants are valued at about 1/5th of their book value—if indeed one can say there is a market value for nuclear power plants at all. In 1999 GPU sold Unit 1 at Three Mile Island, one of the best operating nuclear reactors in the world—a sort of mirror image of its closed and radioactive sister unit—and received virtually nothing for the power plant itself.[46]

Environmentalists were right. Nuclear power could not compete. But having been proven right, environmentalists and other ratepayers were forced to pay the utility to write down the cost of the nuclear power plant so it would be cheap enough to continue to operate. For many anti-nuclear activists it was the most galling of all the policies adopted by state agencies.

Utilities cannot be blamed for not predicting the oil shock of 1973 and the resulting price increases and demand slowdown. But as excess capacity reached historic levels, a prudent management would have delayed or canceled further large power plant construction, including nuclear reactors.[47]

Therefore shareholders should bear the consequences of their managements' decisions.

Since the first nuclear sale in 1998, 13 plants have changed hands. Of the 103 operating nuclear plants in the country, Entergy now owns 14 nuclear plants. Excelon, parent company of AmerGen, controls 21. And recently, to underline just how fast the future changes in electricity these days, the value of nuclear plants had soared, a result of soaring natural gas prices, a looming supply shortage, and increased attention on reducing our reliance on carbon-based emissions from power plants. In 1999 it was a buyer's market in nuclear power, with plants selling for as low as $13 million (Boston Edison's Pilgrim plant). But in 2000 Dominion bid $1.3 billion for the Millstone plant in Connecticut which had years earlier been featured on the cover of *Time* magazine as a poster child for nuclear mismanagement. And the market is so hot that Excelon made a preliminary presentation to the Nuclear Regulatory Commission in early February 2001 about the possibility of building a series of new-technology, small nuclear generators.

The value of nuclear plants recently has soared, and the market is so hot that Excelon made a preliminary presentation to the Nuclear Regulatory Commission in early February 2001 about the possibility of building a series of new-technology, small nuclear generators.

Independent power contracts

While utilities fought long and hard for nuclear power, they fought just as long and just as hard against having to purchase power from independent suppliers. In the early 1980s when several states substantially raised the price utilities had to pay for such power, many utilities sought relief in the courts.

While there is no legal regulatory compact that requires ratepayers to pay off uneconomic nuclear plants, it appears there is a legal requirement that ratepayers pay off the costs of uneconomic independent power contracts. Several state regulatory commissions that tried to change the terms of existing contracts discovered that they lacked the authority to do this.[48] The courts and FERC consistently overturned state decisions in this area.[49]

Just as utility investors received a higher than expected profit on their investments in nuclear power plants, so investors in IPPs received a much higher than expected profit on their investments in plants whose revenue was guaranteed by long-term contracts from utilities.[50]

Some argue persuasively that there were substantial benefits from these IPP contracts. They commercialized new, cleaner technologies. They reduced pollution. They established competition in the electric utility industry. There are some who argue that IPP projects turned out to cost less than the utility plants they avoided. In California, for example, a joint CPUC/ CEC report concluded that these projects were 30-40 percent less expensive than utility projects brought on line at the same time.[51]

As a matter of fairness, one could argue that IPP shareholders shoulder only part of the stranded cost of their high-cost power plants. But as a matter of law, forcing IPP shareholders to pay may be difficult.[52] Regulatory commissions can, however, order their utilities to "mitigate" the costs of the independent power contracts. As of mid 1996, 100 such contracts were renegotiated or canceled, with significant savings to utility customers. It is important to note that many older independent power plants are relatively expensive to run compared to their more efficient successors. Why would an IPP give up a power-purchase contract with high electricity payments? Because in certain situations, a utility is willing to pay an IPP more for not operating the facility.

Rule #9: Where customers are required to pay 100 percent of the excess costs of nuclear power they should have the right to decide whether to continue using nuclear power

Ironically, in states where nuclear power has proven uneconomical, the plants are not being shut down. Rather, customers are being charged 100 percent of the excess costs so the plants can continue to operate competitively.

At the same time nuclear plants are home to an increasing number of temporary storage systems and we have yet to find any community willing to host a permanent radioactive waste site. The federal government is trying to force the states of Nevada and Utah to take all of the nation's radioactive waste on a "temporary" basis, even though those states, ironically, host no nuclear reactors.

Given that one of the key principles guiding our analysis of the new power rules is that we should marry authority and responsibility wherever possible, if customers of nuclear utilities should decide they want to continue generating radioactive waste, their communities should, wherever possible, be responsible for safely managing radioactive waste storage systems.

In those states where customers are required to pay the full uneconomic costs of nuclear power, they should also be given the right to decide whether to continue generating radioactive wastes.

Securitization

Rule #10: Where utilities are paid upfront for stranded costs, regulatory commissions should restrict where they can invest that windfall

Aside from the issue of who pays for stranded costs is the equally thorny issue of how the responsible parties should pay.[53]

Several states have allowed utilities to recover their stranded costs up front.[54] Bonds are issued and paid off by a short-term tax on electricity. Securitization saves ratepayers money because debt costs less than equity; that is, the interest paid on these bonds is less than the return on utility stocks.

But securitization also gives utilities an enormous financial windfall at precisely the moment that utility assets are being reshuffled on a planetary scale. For example, the Philadelphia Electric Company (PECO) will recover $5.26 billion in stranded costs and $4 billion of that will be securitized. Edison International, which serves the Los Angeles market, used over $1 billion in upfront revenue to purchase power plants in New England. In April 1998 the *Wall Street Journal* reported that Texas Utilities (TU) proposed to purchase British Energy Group for $7.4 billion. Coincidentally, TU asked the Texas PUC to force its ratepayers to pay $7.6 billion in stranded costs.[55]

No state has imposed any conditions on how its utilities spend stranded cost money. Some, like Pittsburgh-based Citizen Power, have urged regulatory commissions or state legislatures to require that those stranded costs that are allowed be designated strictly for expenditures that directly benefit local ratepayers. Such expenditures would include, but not be limited to, investments in energy efficiency and renewables.

RULES FOR ACCEPTING RESPONSIBILITY

1. Low income households must have access to electricity
2. Green marketing: favor green citizenship over green consumerism
3. Set an increasing quota for environmentally benign electricity
4. Establish a public benefits fund for renewables and high efficiency distributed power plants
5. Invest in energy efficiency at levels at least as high as expenditures in the peak year (usually 1994)
6. Enact a pollution portfolio standard
7. Substitute pollution taxes for property taxes
8. Stranded costs should be paid by the party primarily responsible for causing those costs
9. Where customers are required to pay 100 percent of the excess costs of nuclear power they should have the right to decide whether to continue using nuclear power
10. Where utilities are paid upfront for stranded costs, regulatory commissions should restrict where they can invest that windfall

CONCLUSION

What Have We Learned? Where Should We Go?

Electricity sprang upon humanity as an almost primeval force, captivating a star-struck public. Its first application—arc lighting—was also its most public, and perhaps its most sensational.

In 1880 the city council of Wabash, Indiana, set up four 3,000 candle-power arc lamps on the courthouse dome. Ten thousand people converged on the Wabash town square on the moonless night of March 31, 1880. When the lights came on, one eyewitness reported, the crowd was "overwhelmed with awe...the strange weird light, exceeded in power only by the sun, rendered the square as light as midday. Men fell on their knees, groans were uttered at the sight and many were dumb with amazement."

Indeed, for many electricity evoked an almost religious response. For Henry Adams, "the dynamo became a symbol of infinity...a moral force, much as the early Christians felt the Cross. Before the end, one begins to pray to it."

From day one electricity was a winner. And with remarkable speed, it expanded its market and reach. Electricity consumption expanded 58-fold from 1902 to 1940. Average household consumption soared from 430 kwhs in 1926 to 4700 in 1964 to over 8000 in 2000. An increasing fraction of all of our energy was used to produce electricity: 10 percent in 1930, 20 percent in 1960, almost 40 percent in 2001. And with the emergence of electric vehicles and the electron-based information economy, the fraction will undoubtedly go even higher.

A century ago, when electricity was just entering society, a fierce debate broke out about its future shape and ownership structure. The debate went on for more than a generation. When the dust finally settled, we had developed rules that embraced a hybrid system: one-third owned directly or indirectly by the electric customer, one-third owned by the investor. Electric companies were given monopoly over both electricity generation and sales. In return for that monopoly and a guaranteed profit, these companies agreed to be regulated by state, and later, federal agencies, and to provide low-cost, reliable electricity to all customers. The electric utility was born.

The regulatory system was largely passive. The regulations adopted were intended largely to channel huge amounts of capital into expanding electric generation and transmission capacity. Demand doubled every ten years, almost as if such increases were an integral part of the natural order.

The future looked both predictable and sanguine. After all, electricity prices had dropped by over 95 percent from 1900 to 1970, when they hovered around a penny a kilowatt hour. System reliability was high. The American electrical system was the envy of the world.

And then the future changed. The oil price hikes and resulting inflation and interest rate hikes in the late 1970s dramatically increased the cost of building power plants while dampening electrical demand. In 1978 federal officials projected electricity consumption to increase by 1990 from 2,100 terawatt hours to 4,100 terawatt hours. Instead, it increased only to 2,800 terawatt hours.

The dramatic hikes in electricity prices, the bankruptcy of several utilities owning nuclear plants, and the inability of the system to stop new, unneeded plants from coming online, galvanized state and federal agencies to become assertive. We changed the rules.

At the federal level, Congress abolished the century-old monopoly utilities had held over electricity sales. That spurred the creation of a new industry—independent power producers.

State agencies assumed a more aggressive stance, becoming in some states full partners with utilities in electricity planning. States insisted that utilities invest in improving efficiency when that saved electricity at a lower cost than building new capacity. By the early 1990s spending on energy efficiency by utilities had gone from nearly zero to almost $3 billion. At the same time, states discovered that utilities were not the ideal vehicle for promoting efficiency, since that conflicted with their primary goal of selling electricity. They adopted various strategies to overcome or avoid this conflict.

Several states began to actively encourage renewable energy. In the early 1980s, California jump-started the commercial wind energy industry by establishing standard contracts that incorporated the high prices that utilities were projecting. In the early 1990s, Minnesota and Iowa mandated a specific level of renewable electricity from wind or biomass.

By the early 1990s many states required utilities to use competitive bidding to acquire new capacity, and in several states, utilities were required to use competitive bidding for efficiency investments as well.

Responding to the crisis of the 1980s, policymakers changed the power rules. But even as they did so, powerful forces were at work that led to the orgy of deregulation that swept through the country between 1996 and 2000. Independent power producers, having become dominant in the wholesale electricity market, lobbied heavily to be able to sell at higher prices directly to final customers. Industrial customers, whose rates partially subsidized household customers, lobbied heavily to be allowed to buy their electricity directly from independent power producers.

Even before the nation embraced electricity deregulation, state initiatives were being undermined. In 1995 the federal government overruled California's policy of allowing only clean energy sources to be included in the competitive bidding process. Between 1994 and 1998 energy efficiency spending by utilities dropped by almost half. Both utilities and independent power producers chose not to build or propose new power plants, preferring to wait until deregulation.

Between 1996 and 2000, almost half the states, with two-thirds of the nation's population, dramatically changed the rules governing their electricity systems. The mantra of those supporting deregulation was "customer choice." They insisted that deregulating electricity was like deregulating trucking. But it is not. Electricity is a profoundly different product from peaches and computers and automobiles. It cannot be sent directly from a power plant to a customer. It has characteristics that interact with equipment and distribution lines. The multiple feedback loops involved in electricity generation and distribution and consumption makes the entire system take on the characteristic of a pulsing, living entity.

The analogy of electricity to peaches was misinformed. The refrain of "customer choice" was misdirecting. It offered customers a choice of suppliers but not a choice of electricity futures.

The price hikes in Illinois and California in the summer of 1999 and 2000, and the rolling blackouts and price hikes of California in the winter of 2001, have revived and energized the deregulation debate. We should take this opportunity to conduct that debate properly this time.

That means approaching electricity as a system. And asking the question, "What kind of electricity system do we want?" We can design many kinds of electricity generation and delivery and storage systems that offer us high reliability and quality and low prices. We should choose the one that offers us as well peace of mind, security and self-reliance.

This book offers a framework for designing the new power rules—to promote authority, responsibility and capacity at the individual and community level. We recommend an electricity system that lessens the distance between those who make the decisions and those who feel the impact of those decisions, and in which we become responsible for our consumption habits.

That means encouraging decentralized power plants. Does this mean autarchic households? Probably not, unless there is a major breakthrough in energy storage systems. Does it mean self-reliant households? Yes. In fact, an intriguing study done by MIT back in the late 1970s concluded that the optimum configuration of a house boasting a solar electric roof is one in which the household exports to the grid half of what it produces and imports from the grid half of what it consumes. Symbiosis, not self-sufficiency, may be the keyword for our electrical future.

The new rules should promote a decentralization of authority as well. That means battling the current trend toward ever more remote economic and political power. It means encouraging customer participation or ownership in the transmission and distribution lines. The empirical data is clear. Customer-owned distribution systems are as reliable and inexpensive as investor-owned distribution systems.

Finally, the new rules should promote a decentralization of responsibility. We should be responsible, in our households and our communities, for the impact of our consumption habits on this generation and the next generation. That means embracing environmentally benign technologies. It also means accepting responsibility for our own pollution. Consider the case of nuclear power. In the last year there has been an amazing turnabout in the economics of nuclear power. Soaring natural gas prices, and higher efficiencies by existing nuclear reactors, have led a revival of the notion of a nuclear-fueled future. One of the arguments in its favor is that nuclear energy does not release greenhouse gases, at least in the actual generation of electricity. However, nuclear energy does generate radioactive wastes. Those communities that embrace nuclear power should accept the responsibility of accepting the wastes generated from those plants.

The debate over the future shape and scale of our electricity system can be an exciting and instructive one. We are living through a remarkable historical moment. Let's not lose the opportunity. Let the debate begin.

Notes

Introduction

1. Hoecker cites "Misconceptions," *Electric Utility Week*, March 31, 1997, p. 2.

2. "TNP Finds Customers Unusually Reluctant to Try Choice," *Megawatt Daily*, January 21, 1997.

3. California:www.cpuc.ca.gov/static/electric/Direct_Access/DASR.htm; Massachusetts: www.state.ma.us/doer/pub_info/migrate.htm; Pennsylvania: www.oca.state.pa.us/

4. Martin Kushler, "Restructuring and Customer Choice: Vox Populi or Dictum Dictatorium?" *Electricity Journal*, January 1998.

5. Margaret Kriz, "Deregulation Shorts Out," *The National Journal*, August 8, 1998.

Chapter 1

1. Other equipment suppliers are still household names: Allis-Chalmers, Brown (now Brown and Boveri), General Electric (which Edison Electric Illuminating became in 1892) and Westinghouse.

2. A kilowatt is a measure of power. It is equivalent to 1000 watts. A 1 kilowatt power plant that operates for 1 hour generates 1 kWh of electricity. A 100 watt bulb burning for one hour consumes 100 Whs or 1/10 of a kWh of electricity.

3. David Morris, *Self-Reliant Cities* (San Francisco, CA: Sierra Club Books, 1992).

4. Electricity has traditionally been generated by burning fuels. The steam is used to turn the generator but much of the heat is wasted. The efficiency of a power plant is measured in the number of units of heat energy it takes to generate a unit of electricity—called the heat rate—which is measured in BTUs per kWh. The first power plants were frightfully inefficient, using perhaps 30 pounds of coal to generate a kWh of electricity. The Pearl Street Station power plant represented a huge leap in efficiency by burning 10 pounds of coal to generate a kWh. Today's power plants burn the equivalent of a pound of coal per kWh produced.

5. David Morris, *Be Your Own Power Company* (Emmaus, PA: Rodale Press, 1982).

6. *Electrical World and Engineer*, May 1908.

7. To make it even more attractive for industries to buy electricity rather than electrical power plants, Samuel Insull developed a tiered rate structure. The first few kWhs consumed were very costly but as the customer consumed more electricity the price per kWh dropped dramatically. As a result industrial rates were much lower.

8. Alfred E. Kahn, *The Economics of Regulation, Principles and Institutions, Vol. 2* (Cambridge, MA: MIT Press, 1988), 119.

9. Morris, *Self-Reliant Cities.*

10. *Ibid.*

11. Leonard S. Hyman, *America's Electric Utilities: Past, Present and Future* (Arlington, VA: Public Utilities Reports, 1994), p. 9

12. See Richard Rudolph and Scott Ridley, *Power Struggle: The Hundred-Year War Over Electricity* (New York: Harper & Row Publishers, 1986) for excellent analysis from two public power advocates.

13. Morris, *Self-Reliant Cities.*

14. Harold Demsetz, "Why Regulate Utilities?" *Journal of Law and Economics*, April 1968.

15. Federal Power Commission, "National Power Survey: Interim Report," *Power Series No. l.*, 1935.

16. Scott Ridley, *Profile of Power: A History of the People and Events that Have Shaped and Continue to Shape America's Most Critical Industry* (Washington D.C.: American Public Power Association, 1996).

17. Hyman, *America's Electric Utilities: Past, Present and Future*, p. 102

18. In 1927 the U.S. Supreme Court ruled that states could not regulate the price of electricity sold in another state, *Rhode Island Public Utilities Commission v Attleboro Steam and Electric Company.*

19. Hyman, *America's Electric Utilities: Past, Present and Future.*

20. Rudolph and Ridley, *Power Struggle.*

21. Scott Ridley makes the point that the intention of the federal government to uncouple the connection to finance that had so encumbered the holdings companies was not well-achieved. Rudolph and Ridley, *Power Struggle.*

22. *Report of the National Power Policy Committee*, 74th Congress, 1st Session, House Doc. No. 137, March 12, 1935.

23. Rudolph and Ridley, *Power Struggle*, p. 81

24. *The Changing Structure of the Electric Power Industry*, Energy Information Administration, 1996.

25. There are two kinds of electricity cooperatives. One is the distribution cooperative. The other is the generation and transmission (G&T) cooperative. The board of directors of the G&T cooperative is comprised of representatives from the distribution cooperatives.

26. See *Otter Tail Power v United States,* 410 US 366, 375-76 (1973) and dissent at 383-387.

27. Virginia A. Coe, Jeffrey Dasovich and William Meyer, *California's Electric Services Industry: Perspectives on the Past, Strategies for the Future* (San Francisco, CA: California Public Utilities Commission, February 3, 1993).

28. "Are Utilities Obsolete? A Troubled System Faces Radical Change," *Business Week,* May 21, 1984, p. 116. The Atomic Energy Commission estimated that by increasing the size of a nuclear reactor from 600 MW to 1300 MW, the cost would drop by over 20 percent, from $703 per kW to $552 per kW. See US AEC's, Power Plant Capital Costs: CurrentTrends and Sensitivity to Economic Parameters, (Washington, D.C.: USGPO, 1974), p. 43.

29. Morris, *Be Your Own Power Company,* p. 17

30. John C. Zink, "Steam Turbines Power an Industry: A Condensed History of Steam Turbines," *Power Engineering,* August 1996, p. 24.

31. *U.S. Commercial Nuclear Power Historical Perspective: Current Status and Outlook* (Washington, D.C.: Department of Energy, March 1982).

32. Ridley, *Profile of Power, op. cit.*

33. Peter Fox-Penner, *Electric Utility Restructuring: A Guide to the Competitive Era* (Vienna, Va.: Public Utilities Reports, 1997), p. 89.

34. During the closing days of 1975, President Ford acknowledged efforts to convince the Saudi Arabian government to invest $1 billion in the "cash-starved" U.S. electric utility industry. David L. Scott, *Financing the Growth of Electric Utilities* (Praeger 1976).

35. This was a time when electric prices were falling but not as fast as the price of other goods.

36. Coe, Dasovich and Meyer, *California's Electric Services Industry.*

37. Rudolph and Ridley, *Power Struggle.*

38. Coe, Dasovich and Meyer, *California's Electric Services Industry.*

39. "Are Utilities Obsolete? A troubled system faces radical change," *Business Week*, May 21, 1984, p. 116.

40. See a discussion of the stranded cost issue in chapter 5.

41. "Are Utilities Obsolete? A troubled system faces radical change," *Business Week*, May 21, 1984, p. 116.

42. *Ibid.*

43. Tom Alexander, "The Surge to Deregulate Electricity," *Fortune*, July 13, 1981, p. 98.

44. Unfortunately, FERC established a very low level for the amount of heat needed to be harnessed. Thus most PURPA machines captured only 5 percent of the waste heat generated.

45. U.S. District Court, Southern District of Mississippi, Civil Action J-79-0212, *State of Mississippi et al. v FERC et al.*

46. AmerGen paid GPU $23 million for the reactor itself and another $77 million for its fuel. *Northeast Power Report*, December 1, 1999.

47. Morris, *Be Your Own Power Company*, p. 29

48. The price drop grew when, in the mid 1980s, FERC allowed industrial customers to purchase gas directly from the supplier rather than through the owner of the pipelines. Ironically, in 1978 Congress had prohibited the use of natural gas for industrial boilers and power plants except by special permission of the Department of Energy. DOE granted waivers liberally. In 1987 Congress lifted this prohibition.

49. As the owner of one of the first and biggest independent power companies explained, "None of these practices are unique to IPPs—utilities could cut costs using these or other methods as well—and in fact, some do. But because utilities' profits are regulated, they have less incentive to cut costs." Roger F. Naill and William C. Dudley, "IPP Leveraged

Financing: Unfair Advantage?" *Public Utilities Fortnightly*, January 15, 1992, p. 15.

50. Peter Fox-Penner, *Electric Utility Restructuring: A Guide to the Competitive Era*, p. 139.

51. *Financial Impacts of Nonutility Power Purchases on Investor-Owned Electric Utilities*, Energy Information Administration, 1994.

52. Charles M. Studness, "The U.S. Supreme Court and Utility Imprudence," *Public Utilities Fortnightly*, January 15, 1992, p. 28. As of 1995 the Nuclear Energy Institute concludes that the total amount of cost disallowances incurred by owners of nuclear plants as a result of prudence reviews "are on the order of $16 billion." Dave Airozo, "NEI says nuclear is competitive if stranded costs are recovered," *Nucleonics Week*, August 17, 1995, p. 13. Although nuclear power plants bore the brunt of these imprudence disallowances, some regulatory commissions also re-examined investments in other types of power plants. The Montana Public Service Commission, for example, refused to allow Montana Public Power to include a 700 MW coal-fired plant (Colstrip 3) into the rate base. The Montana courts overturned the MPSC decision. Later the MPSC rejected another 700-MW coal-fired unit at Colstrip. Montana Power decided to sell its 210-MW share to other regions at a loss. Bob Rowe and Bob Anderson, "The Regulatory Compact," *Montana Business Quarterly*, October 1993. p. 7.

53. Said one observer, "imprudence may be measured as the extent to which a utility fails to achieve optimality in retrospect, not on the extent to which it failed to exercise reasonable judgment." In December 1991 the United States Supreme Court declined to review the findings of Louisiana Public Service Commission that $1.4 billion of Gulf States Utilities investment in River Bend nuclear plant was imprudent. Gulf States had suspended construction in 1977 but restarted it in February 1979. The LPSC concluded that building a lignite plant was the lower cost option.

54. As we shall see, as a result of court decisions, high-cost power purchase contracts, unlike high cost nuclear power plants, were immune from prudence reviews. See chapter 5 for more extended discussion of this point.

55. Timothy Brennan et al., *A Shock to the System: Restructuring America's Electricity Industry* (Washington: Resources for the Future, 1996), p. 31.

56. Joseph Eto, Charles Goldman and Steven Nadel, *Ratepayer Funded Energy-Efficiency Programs in a Restructured Electric Industry: Issues and Options for Regulators and Legislators*, May 1998.

57. Amory B. Lovins, "Energy Strategy: The Road Not Taken?" *Foreign Affairs*, Fall 1976, p. 65-96.

58. One mechanism proposed by Maine regulatory commissioner David Moskovitz in 1989 would have allowed a utility a profit based on the average annual utility bill of its customers compared to those of a set of other utility companies. David H. Moskovitz, "Cutting the Nation's Electric Bill," Issues in *Science and Technology*, Vol. 5. No 3, Spring 1989, pp. 88-93, noted in Stephen Wiel, "Making Electric Efficiency Profitable," *Public Utilities Fortnightly*, July 6, 1989. As Stephen Wiel noted, "If used extensively, linking profit to the relative performance of utilities would likely evolve into a system of competition among all utilities." The idea was never adopted.

59. *Least Cost Utility Planning Handbook for Public Utility Commissioners Vol. 1.*, National Association of Regulatory Utility Commissioners, October 1988.

60. Anecdotally many believed that 1989 spending doubled from the previous year and was static in most of 1980s. Eto, et al., op. cit.

Chapter 2

1. *GAO Reports,* April 24, 1997.

2. Federal Power Commission, "National Power Survey: Interim Report," *Power Series No. l.*, 1935, p. 243.

3. George J. Stigler and Claire Friedland, "What Can Regulators Regulate?" *Journal of Law and Economics*, 1962.

4. Patrick Mann and Walter Primeaux examined a cross section of electricity rates for 113 utilities in 38 states, 8 of which had elected commissions. Mann, an economist at West Virginia University's Regional Research Institute, observed, "the results confirm a price difference caused by regulator selection method; appointed regulators tend to charge higher prices than elected regulators." "Appointed Regulators Tend to Allow Higher Rates than Elected Ones," *Electric Utility Week*, January 10, 1983, p. 6.

5. In 1950 there were more than 80 such cities. By the mid 1960s, 49 such cities remained. By 1981 this had dropped to 27. Today 22 remain. Today most states ban the right of cities to grant competing franchises, a reversal of the situation a century ago when state legislation was required to allow cities to grant exclusive franchises. The 1969 Washington State statute is typical: "The legislature hereby declares that the duplication of the electric lines and service of public utilities and cooperatives is uneconomical, may create unnecessary hazards to the public safety and discourages investment in permanent underground facilities and is unattractive, and thus is contrary to the public interest and further declares that it is in the public interest for public utilities and cooperatives to enter into agreement for the purpose of avoiding or eliminating such duplication." Revised Code of Washington, Chapter 54.48.020.

6. Walter Primeaux Jr., *Direct Electric Utility Competition: The Natural Monopoly Myth* (Praeger: 1986).

7. TCF, *Electric Power and Government Policy: A Survey of the*

Relationship Between the Government and the Electric Power Industries (New York: TCF, 1948), p. 380.

8. See *Who's Who in Cogeneration and Independent Power*, Utility Data Institute, April 1993.

9. The average price of electricity in California rose from 8 cents to 10 cents per kWh from 1982 to 1990.

10. "Regulation of Electricity Sales for Resale and Transmission Service, Phase I," *Federal Register 23*, (June 4, 1985): 445.

11. *San Diego Union-Tribune*. February 28, 1995.

12. The date was move back to April 1 when the utilities found that their computers were not capable of managing the increasing complexity of the marketing system on time.

13. "Power Rates Slow to Cool," *Los Angeles Times*, August 29, 1999, found in Nancy Rader and Scott Hempling, "Promoting Competitive Electricity Markets Through Community Purchasing: The Role of Municipal Aggregation," American Public Power Association, January 2000, p. 24.

14. *GAO Reports*, April 24, 1997.

15. California is an instructive example of the problem of ascribing price reductions to deregulation. California's state legislature guaranteed its electricity customers a minimum 10 percent reduction in their electric bills after deregulation. It accomplished this by taking out a 10-year, low-cost loan to lower electricity costs sufficiently to cover the guaranteed 4-year price reduction. When the additional loan costs are taken into account, the actual reduction was probably closer to 5 percent. Yet many Californians experts in electricity matters estimated that without deregulation electric bills would have decreased by 15 percent or more during the same time period. Thus deregulation, at least in the short term, has arguably resulted in a price increase in that state. The unprecedented increase in electricity prices in late 2000 in California ended any discussion about how deregulation would reduce prices.

16. Edison Electric Institute, *Finance Department Divestiture Action and Analysis*, April 1998. Also, Energy Information Administration, 1998. In 1995 utilities purchased a total of 1283 billion kWh of wholesale electricity from other utilities and a smaller number, 222 billion kWh, from nonutility generators.

17. When electricity is shipped from the coal fields of Montana to Seattle, sometimes 40 percent of it goes south to California. Alfred F. Mistr Jr. and Everard Munsey, "It's Time for Fundamental Reform of Transmission Pricing," *Public Utilities Fortnightly*, July 1, 1992, p. 13.

18. E. Harris, *Life After 888 and 889*. July 1998. Burns & McDonnell Engineering Company. Kansas City, MO.

19. Eric Hirst and Brendan Kirby, "The ancillary forgotten services issue," *Electric Perspectives,* July/August 1998, p. 22-30.

20. Sheila Slocum, "Money, Power, Gas and the Law," *Oil and Gas Journal*, June 29, 1998, p. 56.

21. Enron owns Portland General Electric, although PGE's generating capacity is only a small portion of Enron's overall electricity sales. As of November 2000, Enron was trying to sell PGE.

22. David Wojick, "Regional Power Markets: Roadblock to Choice?" *Public Utilities Fortnightly*, October 1, 1997, p. 28.

23. Allan Sloan, "Utilities Lick Wounds After Encounter with Sharp Toothed Power Traders," *Washington Post*, July 28, 1998.

24. *Wall Street Journal*, July 9, 1998.

25. If that happens, electricity will become the nation's largest traded commodity by a wide margin. After natural gas was deregulated, the volume of trade rose to about $600 billion, ten times the total retail value of natural gas sold nationwide. Financial trading instruments like futures, forwards, swaps and options allow the same electrons to be bought and sold multiple times before they are actually generated and consumed.

26. *Foster Electric Report,* July 15, 1998.

27. "LG&E shelves merchant marketing division, despite top 10 position in power sales," *The Energy Report*, August 3, 1998.

28. *Ibid.*

29. Howard Buskirk, "Emergency Cap Puts Cal-ISO, FERC at Odds on Ancillary Services," *The Energy Daily*, July 15, 1998.

30. Wallace Roberts, "The Dimming Down of America," *The American Prospect*, September 25 - October 9, pp. 54-57.

31. *Foster Electric Report,* July 15, 1998.

32. *Ibid.*

33. *Ibid.*

34. *Reliability Assessment 1998-2007. The Reliability of Bulk Electric Systems in North America.* North American Electric Reliability Council. September 1998.

35. Kiah Harris, op. cit.

Chapter 3

1. A rough rule of thumb is that one 1 kW of capacity is required for every 100 square feet of space or for every occupant.

2. Indeed, shortly thereafter there could be many more. A major technological revolution is occurring not only in the electrical sector, but in the transportation sector, and the relationship between the two is intimate. The typical automobile engine may generate about 50 kW. And the next generation of automobile engines may be emission-free fuel cells. Several companies expecting to produce fuel cells for the automobile industry, which sells 10 million vehicles a year, have decided that the first market for their "engines" will be as household and business-scaled power plants.

3. Joseph Iannucci, Susan Horgan and James Eyer, "The Distributed Future," *Electric Perspective*, May/June 1998, p. 20-32.

4. This analysis of the economics of decentralized power could also be applied to the economics of energy storage. Energy storage is much more expensive than wholesale electricity. But storage systems can correct generation and transmission line imbalances. They can regulate frequency, manage voltage sags and provide area control to avoid damaging utility or customer equipment from excessive voltage drop or a poor power factor. Today the cost of battery systems, including controls and electronics, typically is $1000 per kW. The industry goal is to bring this down to $400. Yet even high-priced storage may be competitive when all the benefits are factored in. In 1995 there was a spirited competition to buy 2,000 megawatts of pumped storage capacity in the United Kingdom from the National Grid Company. The winner was First Hydro, a division of Edison International, which began to provide ancillary services and capacity reserves to the national network and peak power to the power pool. The sale price exceeded $1500 per kW despite the age of the plants. The plant has operated profitably for several years.

5. *Crains Chicago Business*, April 13, 1998.

6. *Ibid.*

7. Tami Cissna, "Price Spike Response," *Independent Energy*, September 1998, p. 34-35.

8. *Wall Street Journal*, October 27, 1998.

9. *Distributed Generation*, National Renewable Energy Laboratory, September 1999 (http://www.nrel.gov/ncpv/pdfs/23398b.pdf).

10. Tom Starrs, email correspondence.

11. "The Growth of Customer Generation," *Electrical World*, September 1999 / October 1999, p.17.

12. Gerry Runte, "Cost Barriers to Distributed Energy Options: Real or Imagined?" *PMA On-Line Magazine*, December 1998.

13. The Electric Power Research Institute (EPRI) Distributed Resources Target began operating in January 1997. The Gas Research Institute (GRI) Distributed Generation Forum began in early 1997. The Distributed Power Coalition of America (DPCA), an offshoot of the Interstate Natural Gas Association, formed in early 1997.

14. Some who envision basement power plants as ubiquitous as basement furnaces note that homes do not have redundant furnace capacity because the gas company will come out and fix a furnace that goes off in the middle of the winter in a matter of hours or less. If a similar repair and maintenance system were available for power plant owners they might dispense with backup power. However, when the heat is turned off it may be hours before a building cools down. But when the electricity goes off, there is an immediate and comprehensive effect and some devices, like timers, must be reset, a greater inconvenience to the building occupants.

15. Archives of the Texas Public Utility Commission's Project #21220 are available on the web at http://www.puc.state.tx.us/rules/rulemake/21220/21220arc/21220arc.cfm.

16. PUC of Texas, Chapter 25, Electric, pp. 21-22.

17. Michael Davis, "Power play: Small plants generating debate," *The Houston Chronicle*, October 24, 2000.

18. "The Role of Energy Efficiency and Distributed Generation in Grid Planning," California Energy Commission, Report to the Governor and Legislature, April, 2000, p. 20-21.

19. Distributed power generation facilities are those that are smaller than those requiring state or federal agency siting permission, i.e. thermal power plants under 50 MW.

20. The distinguishing characteristic of a power plant compared to a furnace or a water heater is that a power plant exports its product outside the home. Therefore interconnection requirements are important. In the late 1980s the standard-making engineering body did develop ANSI/IEEE Standard

1001-1988 IEEE Guide for Interfacing Dispersed Storage and Generations Facilities with Electric Utility Systems. But this standard may not be compatible with the need to develop standards for millions of household and store-based power plants.

21. A handful of advocates seek to allow the duplication of transmission and distribution systems. Mr. Jerry Taylor of the Cato Institute writes, "Other industries with grids already in place could be tempted into the electricity market...Telephone and cable companies have distribution systems and rights of way to virtually all businesses and residences. Many consumers already have access to gas lines. Water and sewer lines provide the rights of way necessary for electricity service. These service providers conceivably could piggyback power lines on their current rights of way and get into the distribution business." In "Is there a need for electric utility reform?" *USA Today* (Magazine), November 1997, p. 10.

22. In May 1998 FERC found Wisconsin Power & Light, Wisconsin Public Service and Illinois Power guilty of refusing to provide transmission services to non-affiliated companies and instead favoring their own affiliates. (*Public Power Weekly,* June 1, 1998.) For an in-depth description of ways that transmission owners can manipulate the system see "Petition for a Rulemaking on Electric Power Industry Structure and Commercial Practices and Motion to Clarify or Reconsider Certain Open Access Commercial Practices," FERC Docket Rm 95-8-000, http://www.elcon.org:80/docres/index.html.

23. Comment of the Staff of the Bureau of Economics of the Federal Trade Commission before the Public Utilities Commission of Texas, Project Number 17549, June 19, 1998.

24. In California, distribution utilities are allowed to own power plants if they demonstrate that such ownership is "consistent with the public interest and does not confer undue competitive advantage."

25. Davis, *The Houston Chronicle*, October 24, 2000.

26. This rate spiral phenomenon is discussed in further detail in "Policies to Support a Distributed Energy System," by Thomas J. Starrs and Howard J. Wenger, Renewable Energy Policy Project.

27. *Ibid.*

28. The price of transporting electricity may be rising, a result of the unbundling effect of deregulation. Once utilities that had previously included in their rate base both generation and transmission (or distribution) assets became only transmission utilities, they tended to add more costs on this asset than they had before. The result has been an increase in charges. A study by the American Public Power Association concluded, "It is obvious that the unbundling process made the sum of the individual charges greater than the previous all-in transmission rates." Cited in Kiah E. Harris, *Life After 888 and 889* (Kansas City, MO: Burns & McDonnell Engineering Company. July 1998).

29. 1997 CADER Conference.

30. Leonard Hyman and Marija Ilic, "Scarce Resources, Real Business or Threat to Profitability?" *Public Utilities Fortnightly*, October 1, 1997.

31. David E. Wojick, "Regional Power Markets: Roadblock to Choice?" *Public Utilities Fortnightly*, October 1, 1997.

32. When electricity is shipped from the coal fields of Montana to Seattle, sometimes 40 percent of it goes south to California. Alfred F. Mistr Jr. and Everard Munsey, "It's Time for Fundamental Reform of Transmission Pricing," *Public Utilities Fortnightly*, July 1, 1992, p. 13.

33. Texas DG comments

34. "Massey: FERC Tilts Toward LMP Model," *The Electricity Daily*, October 3, 2000.

35. Thomas J. Starrs and Howard J. Wenger, "Policies to Support a Distributed Energy System," Renewable Energy Policy Project.

36. *The Role of Energy Efficiency and Distributed Generation in Grid Planning*, Report to the Governor and Legislature, California Energy Commission, April, 2000, p. 20.

37. See chapter 5, "Accepting Responsibility," for a more thorough discussion of an ecological tax shift.

38. Letter from David Morse, Transportation and Public Policy Programs branch, to Eric Wong, Chairman of CADER, August 26, 1997.

39. Wisconsin gives retail rate for renewables and avoided cost for nonrenewables.

40. New York requires residential net metered customers to install equipment to automatically isolate the photovoltaic system from the utility system when voltage and frequency deviations occur, and a manual lockable disconnect switch installed by the customer on the outside of the residence and externally accessible by utility personnel. Moreover, if the utility determines that a dedicated transformer is needed to protect the safety and adequacy of electric service to other customers, a customer generator has to pay up to $350 to install the transformer. To help defray the additional interconnection cost for safety and service reliability, the New York net metering law contains an income tax credit for residents of $0.375 per watt of capacity.

41. Vermont's net metering statute allows for net metering for customers with solar cells, wind or fuel cells if the power plant is no more than 15 kW, but establishes a new class of net metering customers called the farm system. A farm system generates energy from anaerobic digestion of agricultural waste produced by farming and can be up to 100 kW in generating capacity.

42. *Ibid.*

Chapter 4

1. In 1969 Congress allowed IOUs to charge their customers for deferred federal income taxes which are rarely paid because of tax loopholes. Thus, in 1994, more than $57 billion was collected and charged but not delivered to federal government. Congress has proposed to strip COUs of the use of tax-exempt bonds to finance facilities that are used to sell or transmit electricity in competitive markets. The American Public Power Association says that without this and other subsidies, the rates of the private power companies in 1994, on average, would have risen 5 percent. Alan H. Richardson, "Definition of competition varies with the players," *Electric Light & Power*, June 1998, p. 36. In 1986 Congress passed a tax reform act, which established that tax exempt bonds could not be used to finance construction of natural gas, or electric generation transmission or distribution facilities if more than 10 percent of output or use of facilities directly or indirectly benefits private parties. Prior to 1986 the percentage was 25 percent.

2. Cooperatives argue that about two-thirds of their customers are residential; the reverse is true for the IOUs. Two-thirds of their electric sales are commercial and industrial. Electric cooperatives serve predominantly rural areas. While investor-owned utilities have an average of 35 customers per mile, electric cooperatives have an average density of fewer than six consumers per mile. Testimony by Glenn English before Congress, *Federal News Service*, March 25, 1998.

3. Jill Hecht Maxwell, "On the Wired Front," *Inc.* Magazine, September 2000, p. 64.

4. *The Bond Buyer*, June 2, 1998.

5. Maxwell, "On the Wired Front," *Inc.*, September 2000.

6. *The Changing Structure of the Electric Power Industry 2000: An Update*, Energy Information Administration, p. 91.

7. Frank Jossi, "Energy co-ops play for power," *Minneapolis-St Paul CityBusiness*, March 27, 1998.

8. The city council is the board of directors in 61 percent of munis, but the size of the utility matters. This is the case for 70 percent of the two-thirds of the utilities that have fewer than 5,000 customers but for only 33 percent for the 100 utilities that have more than 50,000 customers. About 12 percent of customer-owned utilities elected an independent board of directors directly, although this is 24 percent for larger utilities. In 27 percent of the cases the independent board is appointed, half of the time by the mayor, usually with city council approval. The city council appoints the board in about 40 percent of the cases. *1995 Survey of Administrative and Policymaking Organization of Publicly Owned Electric Utilities* (Washington, DC: American Public Power Association).

9. *Electric Utility Week*, May 11, 1998.

10. Christopher Swope, "Power Politics," *Governing Magazine*, July 1998, p. 42.

11. *Public Utilities Fortnightly*, June 15, 1998.

12. Florida Power & Light and Energy Corp., if their merger is approved, would be the biggest utility to date, with 6.3 million customers, 48,000 MW of generating capacity, and $15 billion in combined revenues. Its market capitalization of $16.4 billion would actually rank only second, behind the holding company formed by the merger of Unicom and PECO, which has a $19 billion market cap.

13. Rebecca Smith, "Deregulation of Power to Jolt Buyers, State," *San Jose Mercury News*. Some have called for a halt to further utility mergers. In early 1998 Joel Klein, assistant attorney in the Antitrust Division of the U.S. Department of Justice, recommended a three-year moratorium on electric utility mergers. In April 1998 the American Public Power Association and the National Rural Electric Cooperative Association asked FERC to impose a two-year moratorium on mergers between electric utilities if more than 1 million metered accounts were involved. No action has been taken on the petition to date. On the other hand, many are proposing to reduce or even eliminate federal roadblocks in the way of concentrated ownership.

For example, in 1998 a bill was introduced (S.621) to overturn the Public Utility Holding Company Act (PUHCA), a law enacted in the 1930s to stop concentrated utility power.

14. *The Changing Structure of the Electric Power Industry 2000: An Update*, EIA, p. 91.

15. Bruce Radford, "Score a Deal? 20-Odd Mergers in Search of a Policy," *Public Utilities Fortnightly*, January 15, 1999.

16. See above for a discussion of equally handsome tax benefits awarded to investor-owned utilities.

17. In the early part of this century, the contest between IOUs and munis was heated. Hundreds of systems changed ownership structures. In recent years the situation has stabilized. Since 1987, 10 municipally owned utilities have been sold to IOUs; three others have been sold to rural coops. During the same period, 12 customer-owned utilities have been created in Ohio, New York, Oregon, New Mexico, Utah and California, 8 in areas previously served by IOUs. Also, "New York towns still want own utilities due to a lack of faith in deregulation," *The Energy Report*, April 27, 1998.

18. www.appanet.org/localcontrol

19. "Pennsylvania Township Gets Involved in Power Biz," American Public Power Association, www.appanet.org

20. "Supplemental Direct Access Implementation Activities Report, Statewide Summary," California Public Utilities Commission, October 15, 2000, www.cpuc.ca.gov/static/electric/Direct_Access/DASR.htm

21. "Promoting Competitive Electricity Markets Through Community Purchasing: The Role of Municipal Aggregation," Nancy Rader and Scott Hempling, American Public Power Association, January 2000, p. 36.

22. National Power's chief executive Keith Henry said his company wanted to enter the retail power supply market without "tying up 1.5 billion pounds (UK) in capital in buying a low-return distribution business as well."

23. "Towns face new decisions over power," *The Boston Globe*, July 12, 1998.

24. The Massachusetts Community Franchise Act also law allows communities to administer funds collected through a system of system benefit charges for energy efficiency and renewable energy programs. See chapter 5, "Accepting Responsibility."

25. "Massachusetts-Ohio Delegation Files Landmark Community Choice Bill in Congress as Restricted California Cities Become State's Biggest Green Power Consumers: San Francisco Asks California Legislature for a Community Choice Amendment," Paul Fenn, American Local Power Association, www.local.org

26. www.ohiocitizen.org

27. The structure and size of the ISOs is often dependent on the self-interest of participants. For example, AEP participated over two years in negotiations with 25 other utilities in an effort to develop a regional ISO in the Midwest. Yet the talks broke down, and over the winter the 26 divided into at least two separate blocs, neither of which includes AEP. Meanwhile, Virginia Power and 11 other utilities announced plans for another ISO. But one of its biggest participants has already dropped out. Detroit Edison said it did not want to turn over control of its transmission network until it gets a clearer picture of what utility deregulation will look like in Michigan, and that it may develop a Michigan ISO. AEP was not included in the alliance with Virginia Power.

28. Wallace Roberts, "The Dimming Down of America," *The American Prospect*, September 25-October 9, 2000, pp. 54-57.

29. EIA, *The Changing Structure of the Electric Power Industry 2000: An Update*, p. 67.

30. *Ibid*, p. 69.

31. "Deregulation fuels opposition to transmission lines," *Electric Light & Power*, May, 1996.

32. Ashley Brown, Testimony before the U.S. House Committee on Energy and Commerce, July 15, 1994.

33. Kiah E. Harris, *Life After 888 and 889* (Kansas City, MO: Burns & McDonnell Engineering Company, July 1998).

Chapter 5

1. Information on low-income provisions of restructuring legislation is taken from: Barbara R. Alexander, *Summary of State Electric Restructuring Legislation: Universal Service Provisions.* Winthrop, ME, January 1, 1998.

2. In 1996 these discounts amounted to $106.9 million.

3. Figures for CO_2 from DOE, Energy Information Administration, *Electric Power Annual 1998*, Vol. II, p. 42; figures for Nox and SO_2 from U.S. Environmental Protection Agency, *National Air Quality and Emissions Trends Report*, 1998, EPA 454/R-00-003, March 2000; figures for Hg from EPA, Mercury Study Report to Congress, Vol. 1, "Executive Summary" EPA-452/R-97-003, December 1997, pp. 3-6. All data cited in Larry Parker and John Blodgett, "Electricity Restructuring: The Implications for Air Quality," 98-615, National Council for Science and the Environment, Environment and Natural Resources Policy Division, July 14, 2000.

4. Virinder Singh, "Fact Summary for Resolution on Sustainable Energy and Low-Income and Minority Communities," Renewable Energy Policy Project, 1999.

5. Barbara Farhar, *Energy and the Environment: the Public View.* College Park, MD: The Renewable Energy Policy Project, 1996. Cited in Steven M. Hoffman, *Energy-Efficiency and Renewable Energy in a Restructured Electric System* (Draft), Saint Paul, MN, October 1998.

6. In California, as noted earlier, less than 2 percent of residential customers have switched suppliers. The vast majority of that 2 percent has taken advantage of a statewide credit for

renewable energy purchases that allows green power providers to offer renewable-based electricity at a price below that offered by the three major IOUs. In California green consumerism has proven popular in part because it is subsidized. Green citizenship offers a means of increasing renewable supplies that doesn't rely on the environmental awareness of typically stingy consumers.

7. Nancy Rader. Draft October 1998.

8. Most green pricing programs allow customers to purchase "fractional" amounts of green power, with a minimum purchase of 100 kWhs per month. This keeps their monthly bills from rising substantially.

9. Green-pricing programs impose the costs of clean energy on less than 2 percent of the customers. Many observers wonder why that should be the primary strategy when polls consistently indicate that a majority of the customers are willing to pay more for clean energy. For example, after an extensive customer poll showed that more than 80 percent of Central and South West (CSW) customers would be willing to pay at least $1 a month more for renewable electricity, the utility announced a green-pricing program for three of its operating utilities in Texas, which required participants to pay a minimum of $4.50 a month more.

10. Connecticut, Maine, Massachusetts, Nevada, New Jersey, Pennsylvania, Texas and Wisconsin.

11. Ryan Wiser, Kevin Porter and Steve Clemmer, "Emerging Markets for Renewable Energy: The Role of State Policies During Restructuring," *Electricity Journal*, January 2000.

12. Connecticut's would add 550 MW of renewable electricity capacity; Massachusetts' 1,300 MW. Connecticut has separate goals for so-called Class I and Class II renewable energy sources.

13. Connecticut is the only state to have a sunset date on its funding for renewable energy technologies.

14. Wiser, Porter and Clemmer, "Emerging Markets for Renewable Energy: The Role of State Policies During Restructuring," *Electricity Journal*, January 2000.

15. US GAO, "Changes in Electricity Related R&D Funding," 1996, p. 6., cited in "Power Switch," REPP, 2000.

16. California has split the 4-year $540 million renewables spending into four categories. Forty-five percent is invested in existing technologies, 10 percent in emerging technologies, 30 percent in new technologies and 15 percent is put in a consumer side account. Emerging technologies money is allocated through a customer rebate program, with a cap of $1000 per customer per year. Sixty percent of the $13.5 million distributed annually in this category must be spent for rebates for small photovoltaic (PV) systems (10 kW or less) with an additional 15 percent reserved for PV systems of 100 kW or less. Twenty-five percent goes toward systems of unlimited size. To encourage early implementation, rebates are reduced by 25 percent a year. New technologies receive a production payment of up to 1.5 cents per kWh during the first five years.

17. *Ibid.*

18. EIA, Annual Energy Outlook 2000, December 1999, DOE/EIA-0383.

19. Eric Hirst, *Electric Utilities and Energy Efficiency*, ORNL, 1998.

20. E. Hirst, R. Cavanaugh and P. Miller, *The Future of DSM in a Restructured U.S. Electricity Industry*, 1996.

21. J. Eto, M. Kito, L. Shown and R. Sonnenblick, *Where did the money go? The Cost and Performance of the Largest Commercial Sector DSM programs* (Berkeley CA: Lawrence Berkeley National Laboratory, 1995).

22. *Unplugged* (Washington, DC: World Wildlife Federation and Environmental Working Group, 1998).

23. Steven Nadel, Toru Kubo and Howard Geller, "State

Scorecard on Utility Energy Efficiency Programs," American Center for an Energy Efficient Economy (ACEEE) April 2000.

24. Richard F. Hirsh and Adam H. Serchuk, "Power Switch: Will the Restructured Electric Utility Help the Environment?" *Environment Magazine*, 1999.

25. Ralph Cavanaugh and Richard Sonstelie, "Energy Distribution Monopolies: A vision for the next century," *Electricity Journal*, September 1998.

26. John Coequyt and Rebecca Stanfield, *Up In Smoke: Congress' Failure to Control Emissions from Coal Power Plants*, Environmental Working Group and the Education Fund of U.S. Public Interest Research Group (U.S. PIRG), 1999, p. 1.

27. *Ibid*, p. 12.

28. *Ibid*, p. 1.

29. Coequyt and Stanfield, *Up In Smoke*, Environmental Working Group and the Education Fund of U.S. Public Interest Research Group, 1999, p. 12.

30. Excerpted from the Executive Summary of "Powerful Solutions: Seven Ways to Switch America to Renewable Electricity," Union of Concerned Scientists, 1999 Fair Pollution Rules.

31. *Ibid.*

32. The Regulatory Assistance Project, "Information Disclosure and Labeling for Electricity Sales: Summary for State Legislatures," Consumer Information Disclosure Series, National Council on Competition and the Electric Industry, April 1999.

33. In 1994 IOUs paid about $13.5 billion a year in various taxes. Cooperatives paid around $650 million. Calculating municipal payments is harder because they pay in lieu of taxes or/and provide free or reduced cost services to local governments.

34. A memo on the process for substituting a pollution tax for existing utility taxes was done by John Bailey of the Institute for Local Self-Reliance, Minneapolis, MN. The report focused on Minnesota. www.ilsr.org

35. Irwin M. Stelzer, "Vertically Integrated Utilities: The Regulators' Poison'd Chalice," *Electricity Journal*, April 1997.

36. Estimates of stranded costs vary wildly. Lower estimates place the cost at roughly $10 billion (comments by the American Public Power Association in the Matter of Recovery of Stranded Costs by Public Utilities and Transmitting Utilities, Federal Energy Regulatory Commission, December 6, 1994) to $20 billion (Michael T. Maloney, Robert E. McCormick and Robert D. Sauer, *Customer Choice, Customer Value: An Analysis of Retail Competition in America's Electric Industry*, Washington, DC: Citizens for a Sound Economy Foundation, 1996, p. 55). Industry analysis often fall roughly between $100-200 billion. (See, for example, *Stranded Cost Will Threaten Credit Quality of U.S. Electrics* New York, NY: Moody's Investors Service, August 1995).

37. Customer-owned utilities, by nature of their ownership, do not make a distinction between shareholders and ratepayers. Therefore the stranded costs battle has not been acute among municipally owned and cooperatively owned utilities. However, customer-owned utilities have had to write down their investments where these have proven high cost. In some of these cases, the national taxpayer, rather than the local ratepayer, has picked up part of the tab. A Government Accounting Office report in September 1997 stated, "It is probable that the Rural Utilities Service will have additional loan write-offs and, therefore, that the federal government will incur further losses in the short term from borrowers that RUS management has identified as financially stressed." According to RUS reports, "about $ 10.5 billion of the $ 22.5 billion in G&T debt is owed by 12 financially stressed G&T borrowers. Of these, four borrowers with about $ 7 billion in outstanding debt are in bankruptcy." *Federal Electricity*

Activities—The Federal Government Net Cost and Potential for Future Losses, GAO, 1997.

38. Kenneth Rose, *An Economic and Legal Perspective on Electric Utility Transition Costs* (Columbus, OH: The National Regulatory Research Institute, July 1996), pp. v-vi.

39. All 16 states that have already mandated competition retail allow some form of stranded cost recovery. New Mexico apparently allowed stranded costs for two utilities, but denied them for a third. Eleven allow 100 percent recovery.

40. Case No. 2761 New Mexico Public Utility Commission.

41. NH PUC Docket N. 96-150. *Restructuring New Hampshire's Electric Industry*. Final Plan. February 28, 1997.

42. "New Hampshire Restructuring Stalemate May Be Over as PSNH, State Reach Accord," *Electric Utility Week*, June 21, 1999, p. 3.

43. Adam D. Theirer, "Electricity Deregulation: Separating Fact from Fiction in the Debate Over Stranded Cost Recovery," Heritage Foundation Reports. March 11, 1997.

44. *Public Utilities Fortnightly*, March 15, 1997. Christopher Seiple, Resource Data International. For slightly different numbers see M.D. Yokell, D. Doyle and R. Kopppe, "Stranded Nuclear Asset and What To Do About Them," presentation to DOE-NARUC Electric Forum, Providence, Rhode Island, April 1995.

45. Other stranded costs included long-term fuel contracts, pension and other worker related liabilities, etc.

46. The reported price was $84 million. But some experts believe that this is about the value of the nuclear fuel stored on-site. The power plant itself went for near nothing.

47. Seventy percent of nuclear's stranded costs is in the 34 nuclear units that went into service after 1984. "The case for the recovery of stranded investment thus hinges on whether the electric utilities built these plants in response

to a public need," writes utility expert Charles Studness. Studness points out that in 1973 excess electrical capacity hovered around 20 percent, about the level considered adequate to meet the reliability needs of regional power pools. But then as new ever-larger power plants continued to come on line, peak demand tanked, plunging from a 7.8 percent growth rate in 1973 to 1.6 percent a year later and only 2.2 percent in 1975. The reserve margin rose to 35 percent by 1985.

"As late as 1978 (utility) managements in aggregate were forecasting that electric demand would grow 5.2 percent a year between 1978 and 1987, even though demand had grown only 3.2 percent annually over the previous five years," writes Studness. "Demand actually grew only 2.2 percent between 1978 and 1987."

Studness' credentials for making his retroactive assessment are validated by his having made a similar prospective assessment. In the June 19, 1980 issue of *Public Utilities Fortnightly,* he observed, "The initial emergence of excess capacity must be viewed as a product of unforeseeable events. But almost seven years later such an explanation no longer suffices. Managements have...failed to be aggressive in slowing down and rescheduling construction..." Charles M. Studness, "The flawed case for stranded cost recovery," *Public Utilities Fortnightly,* February 1, 1995.

48. In 1994, for example, the Ninth Circuit Court of Appeals rejected the California PUC's claim that it had authority to regulate IPP contract entitlements and prices. In New Jersey state regulators tried to revise the purchase price of a contract the year after it was negotiated, but the Third Circuit Court of Appeals held they lacked the authority to do this.

FERC had the authority to step in. In its 1983 decision, the U.S. Supreme Court commented that FERCs avoided cost pricing rule was reasonable "at this early stage in the development of PURPA" and this rule is "subject to revision by the Commission (FERC) as it obtains experience with the effects of the rule." The Court's anticipation that there would be revisions was based in part upon Congress' explicit language authorizing periodic revisions of PURPA rules.

Instead of changing the rules, FERC reinforced the courts' decisions. FERC overruled regulatory commissions in Virginia, Pennsylvania, Oklahoma, New York and California who tried to change the provisions of their purchased power contracts. In Pennsylvania, the independent power producer had not yet invested in the construction of the plant and the 33-year contract term exceeded avoided costs by almost $1 billion, yet still FERC refused to allow the state to change the contract.

Finally, in 1995, FERC declared, in a case involving Connecticut, that a state requirement that a utility pay an IPP more than the utility's avoided cost for power violated PURPA and FERC regulations.

49. The Federal Power Act establishes that federal law, not state law, governs all wholesale power sales. "The provisions of this Part shall apply to the transmission of electric energy in interstate commerce and to the sale of electric energy at wholesale in interstate commerce." Section 201(b)(1). All PURPA qualified sales to a utility are considered wholesale transactions and therefore governed by federal, not state law, even if entirely within a state or even when not entering a high-voltage transmission system. Steven Ferrey, "The QF Cost Dilemma: PURPA Enforcement and Deregulation," *Electricity Journal*, March 1997.

50. The 1995 return on equity for publicly traded IPP companies was estimated to be as high as 27 percent, more than double the typical regulated electric utility's allowed rate of return. Paul J. Kaleta, general counsel at NMPC in Syracuse NY. "Reassessment of PURPA Power Purchase Agreements and PUPA Itself," *The Journal of Project Finance*, Spring 1996. Says Jean-Louis Poirier, senior vice president of Hagler Bailly Consulting, "established companies with 10 or 12 plants, including a bad project or two, have an average rate of return of about 14 to 15%." Hay Worenklein, managing director and head of Lehman Brothers global project finance group confirmed this

view in the same article, "Average company returns in the mid teens compare favorably with the return of 11 to 12 percent enjoyed by oil companies." In "Closely Held Independent Power Industry Enjoys Healthy Profits," *Electric Utility Week*, December 25, 1995, p. 16.

51. Personal communication, Nancy Rader, May 12, 1999. See also, Nancy Rader, *PURPA In Perspective: Early Implementation of PURPA Saved Consumers' Dollars Compared to What Utilities Would Otherwise Have Built*, prepared for the American Wind Energy Association. 1995.

52. Bodington explains the arithmetic. If a utility's power-purchase contract requires it to pay the IPP 10 cents/kWh and the IPP has operating costs of 6 cents/kWh, then the IPP makes a 4 cent/kWh profit. If the utility's own cost of producing electricity from a given power plant is 6 cents/kWh, then it benefits the utility to buy out the contract for about 5 cents/kWh and it benefits the IPP to accept that offer because it makes a 1 cent/kWh profit for not running the plant. Jeff Bodington, "PPAs Under Pressure," *Independent Energy*, May 1996.

53. Another key decision regarding stranded costs is the time frame in which they are recovered. California has the shortest payoff period (4 years), and Pennsylvania the longest (9 years). The shorter the time frame (and the higher the stranded costs) the greater the proportion of the total electric bill comprised by the transition charge. Fully 40-45 percent of the California price of electricity consists of this competitive transition cost (CTC).

This makes it difficult for new suppliers to compete because they can compete only on the energy portion of the bill. Thus for example, if a supplier in California offers a 20 percent reduction in the energy portion of the bill, this translates into only a 4 percent reduction on the total bill since so much of the bill, in the short run, consists of paying off stranded costs. As a result very few small customers have switched suppliers in California.

54. States vary in their approach to securitization. Maine is the only state to forbid it. Illinois and Pennsylvania separate securitization into two phases. Utilities receive half upfront and half at a later date. Maine, Massachusetts and Rhode all have true up provisions.

55. *The Great Ratepayer Robbery: How Electric Utilities Are Making Out Like Bandits at the Dawn of Deregulation*, Safe Energy Communication Council. Washington, D.C. 1998.